Manfred Taferner · Ernst Bonek

Wireless Internet Access over GSM and UMTS

Springer
Berlin
Heidelberg
New York
Barcelona
Hong Kong
London
Milan
Paris
Tokyo

Manfred Taferner · Ernst Bonek

Wireless Internet Access over GSM and UMTS

With 136 Figures

 Springer

Dipl.-Ing. Dr. Manfred Taferner
Neunkirchen, Austria

Dipl.-Ing. Dr. Ernst Bonek
Wien, Austria

ISBN 3-540-42551-9 Springer-Verlag Berlin Heidelberg NewYork

Die Deutsche Bibliothek – CIP-Einheitsaufnahme

Taferner, Manfred:
Wireless Internet Access over GSM and UMTS / Manfred Taferner; Ernst Bonek. – Berlin;
Heidelberg; New York; Barcelona; Hong Kong; London; Milan; Paris; Tokyo:
Springer, 2002
ISBN 3-540-42551-9

Springer-Verlag is a company in the BertelsmannSpringer publishing group
http://www.springer.de
© Springer-Verlag Berlin Heidelberg New York 2002
Printed in Germany

Typesetting: Camera-ready copy from authors
Cover-Design: de'blik, Berlin
Printed on acid-free paper   SPIN: 10850392   62/3020/kk   5 4 3 2 1 0

Preface

We dedicate this book to *students of computer science* who want to understand why the radio channel is at the heart of mobile communications, be it data or voice, and to *students of communications engineering* looking for an introduction to the structure of the Internet protocols, and their application to mobile radio. We also welcome the *practicing engineer* working for mobile network operators and infrastructure suppliers, and the *application programmer* looking for the underlying reasons for the problems they encounter in their daily work.

Given that convergence of the information technology and communications worlds has been talked about for almost 20 years, there is still surprisingly little knowledge among specialists of each others' fields. Data communications over radio, notably Internet access, however, requires full understanding of both fields. Internet protocols have been designed for fixed networks. There bit errors are so rare that interrupted transmission almost inevitably means congestion. In contrast, the radio channel constitutes a bottleneck for mobile data communications in that it introduces bit errors even at modest data rates.

This book is based on the doctoral thesis of one of us (M. Taferner). The work leading to this thesis has been generously sponsored by Mobilkom Austria, a highly innovative and successful mobile network operator in one of Europe's most competitive markets (at the time of writing, Austria boasts a mobile phone penetration of 83%). However, we hasten to say that the views put forward in this book are solely ours, and in no way reflect the views of Mobilkom Austria or Telekom Austria. They are not even based on experience with Mobilkom's network alone, as we have measured our data in different Austrian networks.

It is our pleasure to thank Mobilkom Austria for 12 years of cooperation between Telekom/Mobilkom Austria and the Institut für Nachrichtentechnik und Hochfrequenztechnik of Technische Universität Wien. We have appreciated the continual challenge that has deepened our understanding of the actual needs of an operator. It is probably unjust to give special thanks to just a few persons, but we gratefully acknowledge the assistance of Kurt Ehemoser, Thomas Ergoth, and Alexander Schneider in technical matters, and of Werner Wiedermann and Reinhard Kuch (formerly with Telekom Austria) in all matters relating to the cooperation. We also thank our colleagues Klaus Hugl, Klaus Kopsa and Elmar Trojer for numerous expert discussions. Furthermore, our thanks are due to Georg Löffelmann who contributed the simulations of Chap. 5.6, and Günther Pospischil for help in setting up the measurements.

The benefits of the Internet as we perceive them (instant information retrieval, connectivity at will) will not become universally accepted unless the Internet really becomes ubiquitous, i.e. accessible from mobile terminals. So, in a way, the future success of the Internet is intimately tied to that of mobile radio. Although Japan had a slow start with the acceptance of mobile communications, it is now the largest and fastest growing market in mobile data services. It shows the world the path to the mobile Internet. In Europe, the Scandinavian countries and Austria seem to have the best preconditions for a rapid spreading of the mobile Internet.

We hope that both protocol and system designers, and application programmers working on the 3rd and 4th generations of mobile radio will learn from the successes and deficiencies of earlier generations. If this book contributes a little to this goal, then it will have served its purpose well.

Wien,   *Manfred Taferner*
August 2001   *Ernst Bonek*

Contents

1. Introduction

In recent years, *Internet technology* has been one of the major driving forces behind new developments in the area of telecommunications. The volume of Internet data traffic, i.e. packet-switched data, increases rapidly. The percentage of people with Internet access grows exponentially, too. More and more tasks of daily life can be carried out over the Internet. Meanwhile, Internet access has become almost mandatory in business life. In order to reflect this development, network operators make huge efforts to meet the expected growth of the customers' demands. Backbone as well as access networks are being upgraded in continually shorter intervals. Broadband access networks are currently built using technologies like *Digital Subscriber Line* (xDSL), which drive the demands for even more powerful backbone networks.

In parallel with the development of the Internet, mobile networks face a similar trend, at an even higher level. The enormous success of the European second-generation digital mobile radio standard *Global System for Mobile communications* (GSM, [101]) all over the world brings mobile network operators exponential growth of their customers. In many countries, such as Finland and Austria, the number of mobile subscribers has recently exceeded the number of fixed-line subscribers. The mobile penetration rate at the end of August 2001 has gone up to 79% in Finland, and 83% in Austria. At the moment, Luxembourg has the highest market penetration in Europe with almost 89% [100]. Austria together with Italy holds third place in the mobile subscriber penetration ranking in Europe, having surpassed Sweden and Finland recently.

The combination of both developments, the growth of the Internet and the success of mobile networks, suggests another trend: The growing demand for *mobile access to the Internet*. With second-generation digital mobile radio standards, like GSM, it has become feasible for the first time, to access information from the Internet or private and corporate servers. With emerging standards like the *Wireless Application Protocol* (WAP) an additional booster for mobile data traffic is available. The data traffic will surpass the speech traffic in the Japanese mobile telecommunications market in the year 2001, not least due to I-mode, a similar standard to WAP in Japan. Consequently 'Mobile Internet' is the new catchword.

The market penetration of cellular networks and the Internet is different within the European countries. Figure 1.1 shows a chart with the market penetration rates of mobile cellular systems and the Internet in various European countries. Different areas can be identified: Low cellular and low Internet pen-

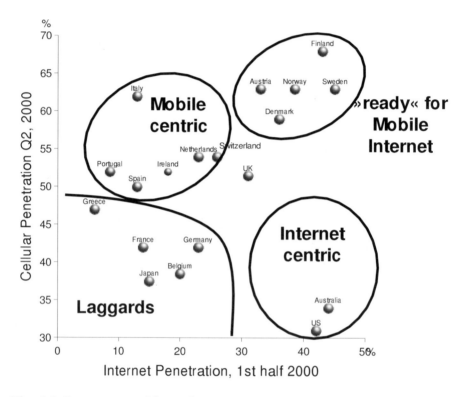

Fig. 1.1. Internet vs. mobile market penetration in European countries in the 1st half of the year 2000 (Source: Ericsson Austria, Nov. 2000)

etration classifies the 'laggards'. In these countries neither mobile networks nor the Internet are used by the mass of the people. [1] High cellular but low Internet market penetration characterize the 'mobile centric' countries, whereas low cellular penetration but high Internet penetration characterize the 'Internet centric' countries. In both classes either cellular systems or the Internet are used by a major section of the people. Best conditions for the establishment of the mobile Internet can be found in countries where both cellular and Internet penetration is high. These countries are shown at the

---

[1] Japan is an exception, because mobile Internet access is already possible to some extent due to I–mode. Despite the subscriber penetration rates being not outstanding there are already many subscribers using the mobile Internet. This must be considered when it is classified as 'laggard'.

right-upper edge in Fig. 1.1. In this area are countries like Finland, Sweden and Norway, the classical leader in cellular market penetration in recent years. But Austria has reached this area recently, which means Austria is ready for the 'Mobile Internet'.

Prior to the GSM Phase 2+, only *circuit-switched data* (CSD) traffic modes have been available. Circuit switched means that a dedicated connection is established for the entire session time, independent of whether actual data is transmitted or not. Therefore circuit-switched lines are *not* adapted to the bursty data traffic from the Internet. No statistical multiplexing gain, which can be expected in view of the burstiness of the traffic pattern, is possible when using a switched line per single connection. Efficient usage of the scarce radio resources is thus almost impossible with circuit-switched data transmission. Also, the data rate available with circuit-switched bearers is not very high, which restricts the application range. For example, in conventional GSM-CSD data rates up to 14.4 kbit/s are possible, together with the *High-Speed Circuit-Switched Data* (HSCSD) extension up to 64 kbit/s are available.

Thus *packet-switched* extensions for the second-generation mobile radio standards have been developed and are now in the starting phase of deployment. Packet-switched data means that data is sent in small portions (packets), and each packet is sent completely independent of each other. No fixed route exists from the sender to the receiver of the packet and no resources are permanently allocated. The packet-switched data support for GSM is the *General Packet Radio Service* (GPRS, [36, 83]). Mobilkom Austria AG has deployed the world's first exhaustive GPRS network during summer 2000. GPRS provides means for a packet-switched data connection from an external packet data network, e.g. the Internet, down the entire path to the mobile station, which includes, in particular, the radio interface. Packet switching on the radio interface is fundamental for efficient utilization of radio resources. The bandwidth is shared between many users. This is possibly because of the bursty nature of data traffic. Idle times of one user are used to transmit traffic of other users (statistical multiplexing). The theoretical data rates in GPRS range from 9.06 kbit/s up to 172 kbit/s, depending on the available radio resource and link quality.

In the upcoming third-generation mobile radio standard, the *Universal Mobile Telecommunication System* (UMTS), packet-switched bearer services have been included in system design from the very beginning. UMTS will provide packet-switched bearer services up to 2 Mbit/s.

The increasing data rates available with GPRS and UMTS, will further drive the amount of wireless data traffic. Entirely new applications become feasible, especially developed for the mobile market. *Location-based services* are just one example. Here the current location of the subscriber determines the application behavior.

Basically, two different methods to access the Internet via a mobile cellular phone exist currently:

- Connection of a radio device to a computer and establishing an *Internet Protocol* (IP) connection in the same manner as a modem connection on the fixed network. The mobile radio device acts basically as a radio modem. In this approach, it is straightforward to utilize the standard Internet protocols as well for transmission over the radio interface. These are the *Internet Protocol* (IP) itself and the *Transmission Control Protocol* (TCP) as transport layer protocol to provide a reliable connection[2]. Use of these well-established protocols is also a question of compatibility to current standard applications and network software in the Internet.
- Usage of special software on the mobile radio phone (e.g. the *Wireless Application Protocol*), to establish a connection via a portal to the Internet. No additional computer is required. WAP in its GPRS version utilizes IP as network layer protocol but has its own higher layer protocol stack.

Both approaches offer advantages and suffer deficiencies. The second approach, dedicated Internet software included in mobile phones, like WAP, is currently becoming a standard in new phones. As the display and computing capabilities of the current mobile phones are still rather limited, the Internet access is only for limited use as well. Not every standard application or Internet web site can be executed or viewed, today. However, the used protocols are optimized for wireless links which could give performance improvements.

The advantages and disadvantages of the first method are pointed out in the following paragraph.

Internet Protocols and Wireless Networks

With the usage of Internet applications on mobile devices, it was a straightforward approach to utilize the *protocols* applied in the Internet also in 'mobile' applications. It was also a question of compatibility to old applications, to use these protocols unaltered. The most important Internet protocols originated in their basic version in the early 1980s. In particular, these are the network layer protocol *Internet Protocol* and the transport layer protocol *Transmission Control Protocol*. IP provides routing and addressing functionality, whereas TCP provides reliable data transmission by means of a go-back-N ARQ mechanism.

Despite continuous enhancements over the years, the basic concept of these protocols remained the same. This concept does *not* provide any mobility functions or special considerations for data transmission over mobile networks.

---

[2] This must *not* be the only layer which provides a reliable connection, e.g. the Radio Link Protocol on the radio interface only in GSM.

A mobility enhancement for IP, called *Mobile IP*, has already been standardized. However, Mobile IP is only suited for global mobility and can not be used efficiently in cellular mobile networks for local mobility management. An equivalent for TCP is not available yet.

The major problem of TCP in wireless environments is the assumption that almost all (99%) of the packet losses are caused by *congestion* in the network. This assumption should be true in a fixed network. However, in a highly unreliable wireless environment, this assumption is *no longer* fulfilled. Hence, the reaction of TCP according to packet losses due to errors on the wireless link is wrong in principle. This has a negative effect on TCP performance.

Additionally, several other protocol mechanisms in TCP sometimes lead to performance losses. This is, for example, the exponentially growing retransmission time in combination with link errors and frequent disconnections. Also the interaction between TCP and link layer protocols in GSM, GPRS and UMTS must not be underestimated.

Several publications deal with the enhancement of TCP to combat the problems outlined above. We will present an overview of suggested solutions in Chap. 5.

Outline

In the second chapter we will present an introduction into the main protocols used for data transmission nowadays. These are, in particular, the data link layer protocols, *Serial Line Internet Protocol* (SLIP) and *Point to Point Protocol* (PPP), and the network layer protocols *Internet Protocol* (IP), *Mobile IP* and *Cellular IP*. Then a detailed introduction into TCP is given. Afterwards a brief overview of the application layer protocol *Hypertext Transport Protocol* (HTTP) is presented, followed by the new *Wireless Application Protocol* (WAP). WAP which is a new global standard to bring Internet content to mobile cellular phones.

Chapters 3 and 4 introduce the basics of data transmission in GSM and UMTS, respectively. We will take special consideration of the protocols involved and their functionality, and the quality of service provided to upper layer protocols.

The essential results for using TCP in a cellular wireless system are given in Chap. 5. This deals with the problems of TCP in wireless environments, and the performance of TCP when used over GSM, GPRS and UMTS. We derive the optimum TCP block length for transparent data transmission. Measurements and simulations of TCP/IP data transmission give a comprehensive performance overview of TCP when used over GSM-CSD, GSM-HSCSD, GPRS and also UMTS. Valuable insights into protocol behavior, pointing out strengths and weaknesses will be given.

Finally, Chap. 6 will draw conclusions from the results obtained in the previous chapters.

2. Protocols for Internet Access

The layered communication reference model introduced by the International Organization for Standardization (ISO), the Open Systems Interconnection (OSI) reference model, serves as basis for almost every communication system. It consist of seven layers, dividing the communication task into separate units. Each layer depends on the services of the layer below in order to provide its own services to the upper layer. Corresponding layers on every connection endpoint are linked by a logical connection, called *protocol*.

The layers, beginning from the lowest level, are: physical layer, data link layer, network layer, transport layer, session layer, presentation layer and application layer.

The physical layer is responsible for the physical transmission of the encoded bit stream, synchronization and access to the physical medium. The physical layer of the GSM system is described in Chap. 3 and that of UMTS-UTRA in Chap. 4. The Internet backbone physical layers are mainly ATM, Frame-Relay or optical transmission systems.

We will describe the next three layers above the physical layer in the following sections in more detail. The functionality of session layer (layer 5) and presentation layer (layer 6) are beyond the scope of this book and we refer to the literature [24]. Finally, tasks and protocols of the application layer are described in Sect. 2.4.

2.1 Data Link Layer

In general, the data link layer takes responsibility for transport of frames across a physical link of the network, for error recovery and link flow control. Two very important data link layer protocols are the *Serial Line Internet Protocol*, SLIP, and the *Point to Point Protocol*, PPP. Both enable IP packet transmission over a serial link but PPP in a more sophisticated way. SLIP and PPP will be described in the following. The data link layer protocols used in GSM-CSD and GSM-GPRS are briefly explained in Chap. 3.

An important parameter of the data link layer is the *Maximum Transmission Unit*, MTU. The MTU limits the amount of data that can be transmitted in a single packet. It is 1500 bytes for the Ethernet standard in LANs. If a connection uses several different data link layers serially one after the other, the smallest MTU is called the *path MTU*. It limits the packet length of the total connection.

2.1.1 Serial Line Internet Protocol, SLIP

The Serial Line Internet Protocol is specified in [117]. It implements a relatively simple mechanism to carry IP datagrams over serial lines. The SLIP protocol is available in almost all standard operating systems.

SLIP uses a simple framing method: Two special byte-wide characters are defined, END (decimal 192) and ESC (decimal 219). A SLIP host simply starts sending the IP datagram and terminates this with the special END character. If the END character occurs in the data stream (in the IP packet) it is replaced by the special two byte sequence ESC_END (decimal: 219, 220). Also, if the ESC character occurs in the data stream it is replaced by the sequence ESC_ESC (decimal: 219, 221). Thus an IP packet can be extracted exactly at the receiver.

Most implementations also send an END character before sending a new IP packet to eliminate possible line noise which would produce corrupt IP packets. But all corrupt packets are thrown away by the IP layer in any case.

An example for framing an IP packet is shown in Fig. 2.1.

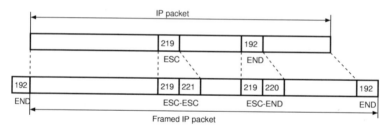

Fig. 2.1. Example of framing an IP packet by the SLIP protocol

Due to its simplicity, some deficiencies exist:

- No addressing: Every host needs to know the IP address of both endpoints.
- No type identification: SLIP can frame only one type of protocol (usually IP). No other protocol can be framed in the same connection.
- No error detection: The error detection or even correction has to be done by the upper layer protocol level.
- Inconstant overhead: The protocol overhead introduced by SLIP is data dependent. Thus in worse cases the overhead could be very high[1].

---

[1] Consider the case of a transmitted image which has very often the color value 219.

A common enhancement of todays implementations is CSLIP (Compressed SLIP), specified in [81] and commonly known as 'Van Jacobson header compression'. CSLIP reduces the protocol overhead especially for small packets. The main idea is to compress the TCP/IP header. Thus the improvements work only for TCP/IP traffic.

Many entries in the TCP/IP header remain constant during a connection. Some of them are just incremented by a constant value, like the sequence and acknowledge numbers of the TCP header. This fact is utilized to compress the 40-byte TCP/IP header[2] to 3 or 5 bytes. The current state of a TCP connection is maintained by each endpoint to enable compression and reconstruction of the TCP/IP header. Up to 16 TCP connections can be handled on the same link by CSLIP.

2.1.2 Point to Point Protocol, PPP

The Point to Point protocol is more sophisticated than SLIP. Most deficiencies of SLIP are corrected by PPP. It consists of the following main parts:

- Framing and encapsulation of network layer datagrams on asynchronous and synchronous serial links.
- A *Link Control Protocol*, LCP, is used to establish, configure, test and close a PPP connection.
- Several *Network Control Protocols*, NCPs, are specified for dynamic configuration of network layer parameters. There exist NCPs for all important network layers, especially for IP (IPCP) and for compression and authentication techniques.

The main parts, framing and encapsulation, and the LCP are specified in [118]. Every NCP is specified in a separate RFC document. For example, the NCP for IP is specified in [98]. The basic frame format is drawn in Fig. 2.2. The format is build after the ISO-HDLC (High-level Data Link Control)

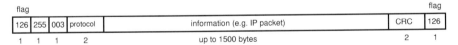

Fig. 2.2. Basic encapsulation and frame format of the point to point protocol (PPP)

protocol. The flag byte delimits a PPP packet. Furthermore, an address byte (always 255) and a control byte (always 003) are included. The protocol field determines the interpretation of the information field. Most important values are listed in Table 2.1. The CRC field holds a cyclic redundancy check,

---

[2] 20-byte TCP and 20-byte IP header, a more detail description of IP and TCP is given in Sect. 2.2 and Sect. 2.3 respectivly.

Table 2.1. Most important values of the protocol field in the PPP frame (IPX is the network layer protocol in Novell networks)

Protocol field value	Contents of information field
0x0021	IP datagram
0x002b	IPX datagram
0xc021	LCP data
0x8021	IPCP data (NCP for IP)
0x802b	IPXCP data (NCP for IPX)

computed over the frame to detect transmission errors.

If the flag byte (126) occurs in the information sequence it has to be escaped (like in SLIP) with the escape byte 125. Byte 126 is transmitted as a two-byte sequence 125–94, the escape byte 125 is transmitted as 125–93.

PPP operation is split into several phases. At first, the *link establishment phase* is used to build up the link by the exchange of LCP 'configure' packets. Next an optional *authentication phase* can be used for authentication of a peer before exchanging network layer packets. After successful authentication, each network layer protocol must be configured separately by the corresponding NCP (e.g. IP is configured with IPCP where dynamic IP address assignment is the most important application). After that, the link reaches the *open state*, and network layer datagrams can be sent over it. The *termination phase* for closing the connection can be entered at any time. Possible causes are loss of carrier, authentication failure, link quality failure, expiration of idle-period timer or administrative closing of the connection. Closing is also done by the exchange of LCP packets.

A number of LCP configuration options exist that are negotiated during link establishment phase. The most important ones are protocol-field-compression and address-and-control-field-compression. Enabled protocol-field-compression reduces the protocol field from 2 to 1 byte. Enabled address-and-control-field-compression omits the first two bytes, the address byte (always 255) and the control byte (always 003). With these two options enabled, the additional overhead compared to SLIP is 3 bytes corresponding to the 1-byte protocol field and the 2-byte CRC field. Additionally, TCP/IP header compression for IP network layer protocol is possible. This is the same as in CSLIP.

PPP implements many features missing in SLIP for the price of a longer link establishment phase, at least 3 bytes additional overhead and a more complex implementation. But one deficiency of SLIP is still remaining: the variable, data-dependent protocol overhead due to and replacing two special bytes, each of them by a special two-byte sequence.

2.2 Network Layer

The most important tasks of the network layer are:

- Transport of packets across a network.
- Unique addressing of end systems.
- Routing of packets through networks, if the endpoints are not directly connected.
- Fragmentation and reassembling of packets.
- Multiplexing of packets on switching points.

Many network layer standards exist for both LANs and WANs, but by far the most important one today is the Internet Protocol, IP. Currently, the most implemented and used version is IPv4. The next section is a short introduction of IPv4. But the next generation Internet Protocol, IPv6, is already specified and we will also provide a brief description of this new version.

Plain IP does not provide any functions for mobility support. *Mobile IP* is an extension to support mobile IP nodes without network address change. Thus every node remains addressable under its unique IP address regardless of any movement. Basic ideas and functions of Mobile IP are explained in Sect. 2.2.3.

A different approach has been introduced with *Cellular IP* [39]. It is a solution to provide IP mobility support in the presence of fast handovers while maintaining the flexibility of IP. Cellular IP is briefly introduced in Sect. 2.2.4.

2.2.1 Internet Protocol, IP

The current version of the Internet Protocol, IPv4, offers a connectionless, unreliable, best effort datagram service, i.e. transportation of datagrams from a sender over various networks to a receiver. It is specified in [109]. IPv4 has no error correction mechanisms included. This task must be handled by upper layers. If an error occurs, the packet is simply discarded. However, the *Internet Control Message Protocol*, ICMP, provides some error detection mechanisms. We refer to the end of this section for further details. 'Connectionless' means that each packet is sent independently and no state information about the connection is maintained. This implies that packets could be sent over different paths, which could lead to out-of-order delivery of the datagrams. IP provides addressing and thereafter mechanisms for routing of datagrams through networks. If necessary, fragmentation and reassembling of large packets is used for transmission through networks with smaller MTU than the original one.

The IP header has a minimum size of 20 bytes, excluding the option field (Fig. 2.3). The header consists of the following parts:

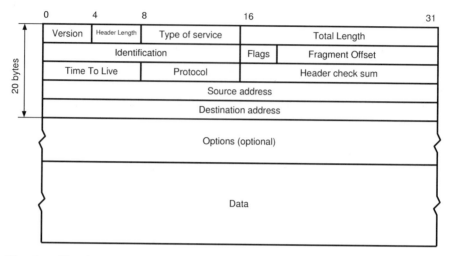

Fig. 2.3. Sketch of an IP datagram including header, options and data portion

- The current *version* of IP is 4.
- The 4-bit field *header length* determines the complete length of the header in 32-bit words. Thus the allowed header length is from 5 (no options) to 60 bytes[3].
- The *type of service* field can be used as an indication of required quality of service of the carried payload. The major choice could be done between normal/low delay, normal/high throughput, normal/high reliability, normal/minimize monetary costs (4 bits) and another 3-bit precedence value, which allows different service precedence.
- *Total length* field contains the total length of the IP datagram in bytes. As it is 16 bits long the maximum IP datagram size is limited to 65 535 bytes.
- An *Identification* is used to uniquely identify each datagram. IP packets containing parts of the same fragmented upper layer protocol packet should have the same *identification* value.
- The *flags* field consists of 3 bits but only 2 bits are actually used. The *Do not fragment bit* indicates a datagram that must not be fragmented, whereas the *more fragment bit* is set when the actual datagram is not the last datagram of a sequence of a fragmented upper layer datagram.
- The *fragmentation offset* field indicates the position in the upper layer datagram this fragment belongs to. The first fragment has offset zero.
- The *time-to-live* field indicates the maximum number of hops a datagram is allowed to pass.
- The *protocol* entry determines the upper layer protocol stored in the data portion of the IP datagram.

---

[3] The 4-bit field limits the header to 15×32-bit length.

- The *header check sum* covers only the header. Since some parts of the header change each node the IP datagram passes, it must be recomputed every time the datagram is processed. The data portion is not included.
- The *source address* and *destination address* are 32-bit values which should be unique for every interface. The IP address is hierarchical and five different address classes exist (class A to E). Additionally, three types of IP address, unicast (a single host), broadcast (all hosts on a given network) and multicast (a set of hosts on a given network) are available.
- The *option* field has variable length. As the total header length is limited by the 'header length' field, the maximum length of *option* field is also limited by it. Options may appear or not. Currently the following options exist: Security and handling restrictions (for military applications), record route (each router records its IP address), timestamp (each router records its IP address and time), loose-source routing (list of addresses that must be passed), strict source routing (list of addresses which are the only allowed addresses to pass).

The IP layer is also responsible for routing datagrams through the network. A number of IP routing protocols exist which implement strategies of routing IP packets to the destination. IP routing is done on a hop-by-hop basis, IP does not know the complete route through the network to a destination. IP routing uses hierarchical addresses extensively.

As already mentioned above, the *Internet Control Message Protocol* (ICMP) [110] can be used to send error messages or other messages for network diagnostics. ICMP packets are sent as IP datagrams. ICMP uses the the basic support of IP to send datagrams as if it were a higher-level protocol, however, ICMP is actually an integral part of IP. It must be implemented in every IP module. The purpose is to provide feedback in the network but *not* to make IP reliable. A node (router or end system) that encounters a transmission problem generates an ICMP message and sends it back to the end system where the datagram, which causes the problem, originates.

2.2.2 Internet Protocol, Version 6, IPv6

IPv6 is a new version of IP which is designed to be an evolutionary step from IPv4. It is defined in [42]. The key issues in designing IPv6 were twofold: First, IPv6 must be able to support a large global Internet and second, there must be a clear way to transition the current large installed base of IPv4 systems.

The changes from IPv4 to IPv6 cover mainly the following categories:

- Expanded routing and addressing capabilities. The address size was increased from 32 to 128 bits. Thus a much greater number of nodes can be addressed, the number of hierarchy levels can be increased and auto-configuration is simplified. Also, a new type of address, an *anycast* address is defined. It is used to send a packet to any one of a group of nodes.

A new address space will enable a significant reduction in the number of routes required to be maintained by the backbone routers. However, this will become only true if the entire Internet switches to IPv6.

- Header format simplification. Some IPv4 header fields have been made optional or even dropped. This makes header processing costs smaller and keeps protocol overhead of IPv6 as low as possible despite the increased size of addresses. The new IPv6 header is shown in Fig. 2.4. Even though

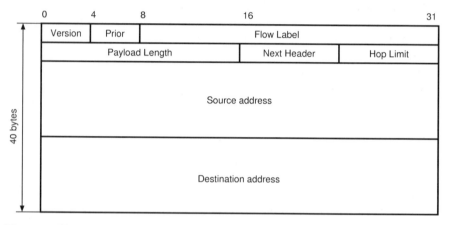

Fig. 2.4. IPv6 header format without extensions and data field

the IPv6 addresses are four times longer than the old IPv4 ones, the IPv6 header is only twice the size of the IPv6 header. *Nevertheless, the protocol overhead has doubled!*

- Improved support for extensions and options. The way options are encoded allows more efficient header processing, more flexibility for adding new options and less stringent limits on the options length.
- Improved Quality-of-Service capability. Labelling of packets belonging to particular traffic flows for which the source requests special handling is improved. This can be used for non-default quality of service or even real-time service.
- Authentication and privacy capabilities. IPv6 includes definitions for extensions which provide support for authentication, data integrity, and confidentiality. This is included as a basic element of IPv6.

An important point in IPv6 considers packet size issues. IPv6 *requires* that every link in the Internet has an MTU of at least 1280 bytes. On links that are not able to transmit such packets, link-specific fragmentation and reassembling must be provided in the layer below IPv6. Configurable link layer protocols, such as SLIP or PPP must be configured to have an MTU of at least 1280 bytes.

2.2.3 Mobile IP

Mobile IP provides extension to IP to enable node mobility in the Internet. It enables routing of IP packets to mobile nodes which may be connected to any link while using their permanent IP address. It is specified in a number of IETF RFC documents. The Mobile IP protocol itself is defined in [102]. Besides this, there are definitions for tunneling mechanisms in [76,103,104], a document which describes the applicability of Mobile IP [119] and a specification of the Mobile IP Management Information Base (MIB) [40]. The Management Information Base (MIB) is a collection of variables in every Mobile IP node, which can be examined or modified by a manager station using the Simple Network Management Protocol version 2 (SNMPv2) [38]. Tunneling is the encapsulation of an entire packet in the payload of another packet between an entry and an exit point of the 'tunnel'. The decapsulation is done at the end of the tunnel, thus getting back the old IP packet at a different location in the network without modifying the packet itself.

The three basic functional entities in the Mobile IP protocol are drawn in Fig. 2.5. A *Mobile Node* is a node that can change its point-of-attachment

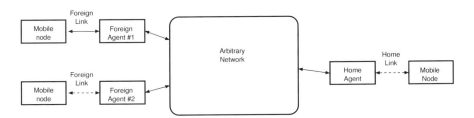

Fig. 2.5. Overview of Mobile IP entities and relationships

to the Internet. It is either attached via its home agent or a foreign agent. The *Home Agent* is a router[4], which holds the *care-of-address* of a mobile node. It intercepts a packet destined to the mobile node and tunnels it to the mobile node's current address. The *care-of-address* is the current address of the mobile node in a foreign network. The *Foreign Agent* is a router which detunnels packets coming from the home agent and forwards it to the mobile node. A *home address* is permanently assigned to every mobile node, which is the address the mobile node can be addressed with. The home address is fixed. If the mobile node is currently attached at its home agent via the home link the home address is used to route packets to the mobile node.

If the mobile node is moving to another point-of-attachment, e.g. to foreign agent #1, the following procedure will occur. An illustration of this procedure and also a packet flow example is drawn in Fig. 2.6.

---

[4] In this context a *router* is a network device that forwards IP packets not destined to itself.

- Every foreign agent and also every home agent periodically sends *Agent Advertisement* messages. These are either multicast or broadcast messages.
- The mobile nodes detect these messages, so they can determine whether they are connected to the home agent or any other foreign agent. If a mobile node is connected to its home agent it acts like a fixed host and no further Mobile IP functionality is used. It is addressed under its home address (case 1 in Fig. 2.6).
- If a mobile node determines that it is connected to a foreign agent it acquires a care-of-address (case 2 in Fig. 2.6). There exist two different kinds of care-of-address: The *foreign agent* care-of-address is the IP address of the foreign agent the mobile node is connected to. The *collocated* care-of-address is the IP address temporily assigned to the mobile mode while connected to the foreign agent. These two kinds of care-of-address are assigned to the mobile node by the foreign agent.
- The mobile node has to register its care-of-address at its home agent by a special authenticated registration procedure. Thus the home agent is aware of the mobile nodes new address. The registration has to be authenticated to prevent other users from registration of bogus care-of-addresses, otherwise any traffic to the mobile node could be disrupted.
- The home agent still receives all packets destined to the mobile node. But it knows now the current point-of-attachment of the mobile node and tunnels the packets to the foreign agent care-of-address. The foreign agent decapsulates the received packets and sends it to the mobile node via its temporily assigned collocated care-of-address.
- In the reverse direction, the packets sent by the mobile node to any destination are routed directly to the destination via the foreign agent, without the need of any tunneling or the help of any care-of-address.

The routing to and from a mobile node attached to a foreign agent is also called 'triangle routing'. This is because every packet destined to the mobile node is routed via its home agent whereas the packets outgoing from the mobile node are directly routed to its destination via any roundabout way. To route the packets more efficiently to the mobile host in certain situations, a proposal for route optimization exists (e.g. [105]).

Mobile IP does not require any change in hardware or software in the existing base of installed hosts and routers besides the entities involved in the Mobile IP procedures. Thus Mobile IP can be installed only in the entities needed, there is no need for a global upgrade of all other network parts.

Every time the mobile node changes its point-of-attachment, the home agent must be informed of this change, a new registration is required. This is called a *location update*.

For a more detail introduction into Mobile IP see [106, 120].

Mobile IP is a well-established solution to provide mobility to mobile hosts. However, it introduces significant overhead due to registration proce-

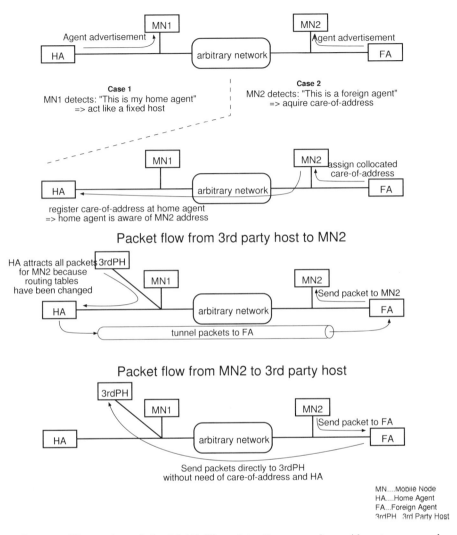

Fig. 2.6. Illustration of the MobileIP registration procedure. Also, two examples of a packet flow from and to a mobile node and a 3rd party host

dures and triangle routing. It is not well suited for frequent location updates, but only for wide-area routing.

2.2.4 Cellular IP

When the number of location changes of a mobile node increases, the signalling traffic of the Mobile IP protocol could be a significant amount of the total traffic. This is especially the case if Mobile IP is used in cellular networks with frequent handover from one access point to another. A new approach to deal with this problem is proposed in [37, 128].

The basic idea is to *separate local and wide area mobility*. For wide-area mobility (i.e. relatively infrequent changes of the access network) the existing Mobile IP protocol is used. Mobile IP supports mobility with a granularity of the access network. For local mobility (i.e. mobility *within* the access network) the new *Cellular IP* protocol is introduced.

Cellular IP maintains a distributed cache for location management and routing purposes. When a mobile node attaches an access network it must send paging update packet's to the gateway router[5]. These packets are examined by every node in the access network on the packets way to the gateway router to determine the route to the mobile node. Every node then creates 'mappings', which is basically a local routing table for every node. Thus the access network is now able to route an incoming packet to the mobile node. Update mechanisms are used to keep the information up to date when the mobile node is idle or moves to another access point. Also, every access point periodically emits beacon signals to allow for mobile nodes to identify the current base station they are attached with. The mobile nodes do not have to register at any agent in the case of a change in their point-of-attachment. A handover does not lead to additional traffic. However, it is possible that packets get lost during a handover.

Because the system is fully distributed, it leads to a simple and lightweight implementation. Cellular IP supports frequent handovers between access points, which is a major disadvantage in Mobile IP. Additionally, neither new packet formats are required nor address space allocation, beyond what is present in IP, is needed.

2.3 Transport Layer

The transport layer is responsible for end-to-end connections, which could be reliable or unreliable. It uses the services of the lower network layer to maintain a flow of data between the two connection endpoints. Also addressing of processes inside the end system is accomplished.

In the TCP/IP protocol suite there exist two transport layer protocols: The *Transmission Control Protocol*, TCP, and the *User Datagram Protocol*, UDP. TCP is by far the most important reliable transport layer protocol in today's networks. We will describe both TCP and UDP in more detail in the following sections. A comprehensive description of TCP can be found in [124].

---

[5] The gateway router links the access network to the global Internet.

2.3.1 Transmission Control Protocol, TCP

TCP provides a reliable, connection-oriented service between two hosts. The original specification of TCP is [111]. However, a number of changes and enhancements have been introduced since this original specification. The most widely used versions of TCP today are 'TCP Tahoe', which was the first one and 'TCP Reno'[6]. The Reno modifications are specified best in [22]. Recently a new version, TCP Vegas, has been published [32] which we briefly describe at the end of this section.

TCP achieves reliability by the following main mechanisms:

- The receiver sends acknowledgment packets, so the transmitter is aware of the correctly transmitted segments.
- Each portion of transmitted data, called a *segment*, is secured by a check sum. Thus transmission errors in each segment can be detected.
- When a segment is sent, a timer is started. The sender waits for acknowledgment for each segment, until this timer goes off. Then this segment is retransmitted.
- If segments do arrive at the receiver in the wrong order[7], TCP sends those segments reordered to the application.
- TCP discards duplicate received segments.
- And TCP provides *flow control* mechanisms to match the sending rate to the capacity of the pipe and the hosts.

TCP also relies on the important assumption that packet loss due to transmission errors is very small (much less than 1%). Thus most of the segments that are lost in the network are lost due to congestion, i.e. overflow in network elements (e.g. buffers) in the network.

TCP header
The TCP header format is drawn in Fig. 2.7. The identification of the application the transmitted byte stream belongs to is made by *port numbers*. The TCP header contains a source and a destination port number. A port number together with an IP address determines a so-called *socket* that identifies a unique application at a specific host.

The *sequence number* is the byte number of the first byte of the actual transmitted segment in the total flow of the transmitted byte stream. Thus each segment is identified in the entire transmitted byte stream.

The *acknowledgment number* in an acknowledgment packet contains the next sequence number the receiver expects to receive. An acknowledgment packet is signed by the *ACK* (A) flag set in the header. Among the six flag

---

[6] The TCP versions are named according their first appearance in the Berkeley Software Distribution, BSD Unix.

[7] The lower network layer does not take responsibility for in-order delivery of datagrams.

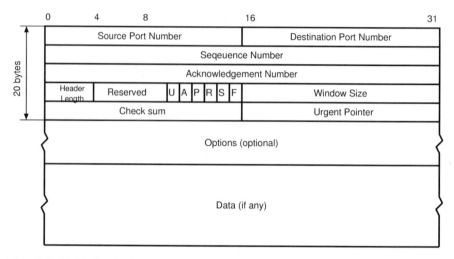

Fig. 2.7. TCP Header format including options and data

bits, *SYN* (S) and *FIN* (F) flags are used for connection establishment and connection release.

To determine the actual header length inclusively all options the *header length* field is used. It contains the header length in 32-bit words.

TCP includes flow control mechanisms. For this a sliding window is advertised by the receiver which is contained in the *window size* field. Added to the acknowledgment number this determines the maximum sequence number the receiver is willing to accept.

The *TCP check sum* is the check sum over the header including options *and* data of the segment. Also an IP pseudo-header with IP source and destination address is included. This makes it possible to check the entire TCP/IP address with the TCP check sum.

Connection establishment

As TCP is a connection-oriented protocol, a connection must be established at the beginning of the data transmission and terminated at the end. A *three-way handshake* is used for connection establishment. This determines unique initial values for the sequence and acknowledgment numbers. Thus a unique synchronization point for the start of the data transmission phase is guaranteed regardless any old TCP segment still traveling through the network. The handshake uses the SYN and ACK flags.

The initiating host sends a segment with its own sequence number together with the SYN flag set. Thus the partner host knows the acknowledgment number it has to send. It responds with a segment including its own sequence number, the ack number just got from the initiating host together with the SYN and ACK flag set. The initiating host is now also aware of the sequence number the partner host is using, thus knowing how to acknowledge packets coming from this host. To complete the handshake, the initiating host sends an acknowledgment packet back to the partner host. Now the connection is opened for data transfer in both directions.

Data transmission
TCP is an ack- or self-clocked protocol. In the stationary state, the incoming rate of the acknowledgment segments determines the sending rate of the data segments (Fig. 2.8). With this mechanism, the data flow is automatically

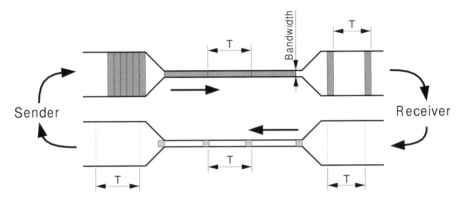

Fig. 2.8. TCP self-clocking mechanism. The rate of the incoming acknowledgment segments (lower level flow) determines the sending rate of the data packets (upper level flow). Thus the flow is automatically adopted to the lowest bandwidth in the pipe. Bandwidth is sketched as vertical dimension

adjusted to the bandwidth of the bottleneck in the pipe (the middle section in Fig. 2.8). This is the basic flow control mechanism in TCP.

Every sent segment is explicitly acknowledged by an acknowledgment segment. Acknowledgments are cumulative, i.e. an acknowledgment with a higher sequence number acknowledges all older packets as well. If an error occurs in some data segment the receiver waits with the sending of the acknowledgment until the missing segment is received. However, *every* received data segment causes an acknowledgment segment, but in the case of an error, the acknowledgment number remains the same as for the last correctly received data segment (duplicate acknowledgments). Every segment is secured by a timer started at the transmission time. If the acknowledgment is not received by the sender until this timer goes off (retransmission timeout) the segment is retransmitted. This is the only way for error correction in TCP. The retransmission timeout is crucial for performance. It should correlate with the round trip time, RTT. If it is too high in relation to RTT it means too long a delay in the case of a retransmission. If it is too low, unnecessary retransmissions occur. The timeout (RTO) is computed by the following al-

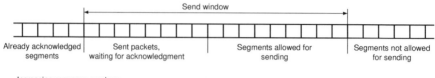

Fig. 2.9. TCP sliding window mechanism. Every packet is stored in a buffer and the state of the packet is maintained. All segments between the last not yet acknowledged segment and the last not yet sent segment lie inside the sliding window. The window moves towards increasing time during data transfer

gorithm [79, 80][8]:

$$\text{Err} = \text{RTT}_{\text{measure}} - A$$
$$A = A + g \cdot \text{Err}$$
$$D = D + h(|\text{Err}| - D)$$
$$\text{RTO} = A + 4D . \tag{2.1}$$

$\text{RTT}_{\text{measure}}$ denotes the actual measured round trip time of the last received segment, A is an estimator of the average RTT, D the smoothed mean deviation, g a gain factor for the average computation, typically set to $g = 1/8$ and $h = 0.25$ the gain for the deviation. This computation can be done in integer arithmetic [79], thus the values of g and h are all powers of 2. But this algorithm must not be executed for received acknowledgments where retransmission has occurred. Otherwise the RTT estimate would be wrong. This is known as 'Karn's algorithm' [84]. If a retransmission occurs, the RTO for this segment is exponentially backed off, which means it is doubled for every retransmission of the same segment. This exponential back-off time is necessary to provide stability of the protocol.

Sliding window mechanism

TCP implements a *continuous repeat request go-back-N* (CRQ-GBN) protocol. This means, if an error occurs, all packets starting from the packet in error have to be retransmitted[9]. It maintains a buffer where the actual state of each segment is stored. The segments are identified by their sequence numbers. This buffer is illustrated in Fig. 2.9. Every segment is in one of the following four states:

1. Sent and acknowledged.
2. Sent and *not* acknowledged.
3. Allowed for sending but not yet sent.

---

[8] This is not the original algorithm specified in [111]. The author of [79] pointed out that the original algorithm cannot keep up with wide spreading of the RTT. This could cause unnecessary retransmissions.

[9] This behavior can be optimized by the TCP selective acknowledgment option [97]. With this option only the packet in error has to be retransmitted.

4. *Not* allowed for sending.

Those segments belonging to the states 2 and 3 mark the *sliding window*[10]. During data transmission, the window moves towards increasing sequence numbers. The size of the window determines most of the dynamic TCP protocol operations and is thus an essential mechanism of the TCP protocol. The performance is heavily determined by the window behavior. In the implementations, the sliding window is computed as the minimum of the *congestion window* and the *receiver window*. The congestion window is maintained in the transmitter and is determined by various dynamic features of TCP (see below). The receiver window is the window advertised by the TCP receiver to the TCP sender. Thus the receiver has an additional method (besides the ack-clocking behavior) to influence the TCP sender's sending rate. At any given time, a TCP sender must not send data with a sequence number above the sum of the highest acknowledged sequence number and the sliding window. If a packet is acknowledged by the receiver and the size of the sliding window remains unchanged, the sliding window moves a single segment towards higher sequence numbers thus allowing the sender to send the next segment at the higher end of the sliding window.

Slow start

At the beginning of a data transmission over a network which has an unknown topology, the sender does not know the maximum rate it can send out segments without producing network congestion. Thus the sender has to probe the network for the highest possible sending rate. TCP provides the *slow start* mechanism to accomplish this probing in a first step.

At the beginning after the three-way handshake, the congestion window, cwnd, is set to one segment[11], $cwnd_0 = MSS$. Thus the sender is allowed to send initially just a single segment. Every time an acknowledgment arrives at the transmitter for a sent segment, the window is increased by one segment, $cwnd_{i+1} = cwnd_i + MSS$ (see Fig. 2.10). This leads to an exponential increase of the congestion window in time. The exponential increase is illustrated in the left part of Fig. 2.11.

Congestion avoidance

If the slow start phase is continued above a point, where the sender transmits the segments at a higher rate than the maximum possible rate determined by the bottleneck link in the path, *congestion* in the network will appear. That is, the buffer of the network element located before the bottleneck link will be filled up and thus eventually segments will be discarded. To avoid this condition, the *congestion avoidance algorithm* is used.

Two indications for congestion in the network exist: a timeout occurring and a receipt of a duplicate acknowledgment.

---

[10] The sliding window is often called *send window*.
[11] MSS denotes the Maximum Segment Size in bytes. The congestion window is maintained in bytes, but it is always changed in units of segments.

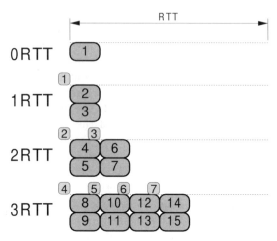

Fig. 2.10. Illustration of the slow start phase in TCP (from [79]). The large rectangles mark the data segments sent, the small ones mark the received acknowledgment segments. The time line goes from left to right and subsequently from up to down. Each line corresponds to one RTT. Segment 1 is sent at the beginning and it takes 1 RTT until the corresponding acknowledgment is received. Then the transmitter is allowed to send segments 2 and 3 because the congestion window has been incremented by 1 (slow start algorithm). After a second RTT the acknowledgments 2 and 3 arrive which themselves trigger the sending of data segments 4 to 7. Thus the number of segments transmitted per RTT increase exponentially

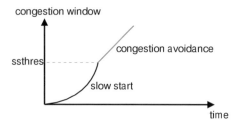

Fig. 2.11. Increase of the congestion window over time in the slow start and congestion avoidance phase in the beginning of TCP data transmission

Congestion avoidance is always implemented together with slow start. It introduces a new variable, the slow start threshold size, ssthres. If congestion occurs in slow start phase (a timeout or duplicate ACK has occurred), the current congestion window size is saved in ssthres. At the same time, the slow start phase ends and the congestion avoidance phase begins, where now the congestion window is increased only by a single segment in *one* RTT. This means the congestion window is increased by $cwnd_{i+1} = cwnd_i + MSS/cwnd_i$ every time an acknowledgment arrives. If congestion is indicated by a timeout, slow start phase is entered again. However, the old slow start threshold value has changed, meaning that the next starting point for the congestion avoidance algorithm has changed.

The congestion avoidance algorithm leads to a linear increase of the congestion window in time, indicated in the right part of Fig. 2.11.

Fast retransmit

Until now a packet is only retransmitted if the corresponding retransmission timeout occurs. If the connection is running over a pipe having a long round trip time, the retransmission timeout value could be very large. But if acknowledgments still arrive at the sender, we know that the path is still opened, just a single segment is missing. The acknowledgments arriving at this time could only be duplicate acknowledgments, because only the last correctly received segment is acknowledged. So we know, if duplicate acknowledgments arrive, there is possibly just a single segment missing.

So the *fast retransmit* algorithm performs an immediate retransmission of the oldest not-yet-acknowledged segment when three duplicate acknowledgments arrive at the transmitter.

In the TCP Tahoe version, fast retransmit was immediately followed by the slow start phase. In the newer versions (TCP Reno and Vegas) it is followed by the fast recovery algorithm.

Fast recovery

Because the reception of duplicate acknowledgments indicates that data is still flowing through the network, it is inefficient to follow a new slow start phase after the fast retransmission of the lost segment.

Instead we set the slow start threshold value, *ssthres*, to one-half of the current congestion window[12] and reduce the congestion window to ssthres immediately after an acknowledgment arrives that acknowledges new data (no longer duplicate acknowledgment). The behavior of the congestion window in the case of fast retransmit and fast recovery is plotted in Fig. 2.12.

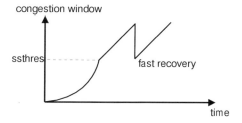

Fig. 2.12. Increase of the congestion window over time in the slow start and congestion avoidance phase in the beginning of TCP data transmission. After reception of duplicate acknowledgments the fast recovery algorithm is started

---

[12] Strictly it should be one-half of the minimum of the advertised window of the receiver and the current congestion window.

After the congestion window is set to *ssthres* immediately the congestion avoidance phase is entered (linear increasing of the congestion window over time) instead of the slow start phase.

TCP Reno

The most important differences of the 'Reno' version to the above described 'Tahoe' version of TCP are summarized in the following:

- The initial congestion window after the three-way handshake at the beginning of the slow start phase should be set to $cwnd_0 \leq 2 \times SMSS$, where SMSS is the maximum segment size the TCP sender is allowed to transmit.
- Congestion avoidance phase now increases the congestion window by $cwnd = cwnd + SMSS \times SMSS/cwnd$, which is a slightly more aggressive policy, because congestion window is increased slightly more than one full-sized segment per RTT.
- The fast retransmission algorithm is now followed by the *fast recovery* algorithm (see above for description).
- In fast recovery, the ssthres value must be set to $ssthres = \max(FlightSize/2, 2 \times SMSS)$. FlightSize is the amount of data that has been sent but not yet acknowledged.

Additionally, some other modifications regarding restarting connections or sending delayed acknowledgments have been introduced. For those we refer to [22].

TCP Vegas

TCP Vegas has been proposed in [32] to provide between 37 and 71% better throughput, with one-fifth to one-half the losses as compared to the TCP Reno version. TCP Vegas does not involve any changes in the standard but is an alternative implementation that interoperates with any other valid TCP implementation. All changes are only confined to the sending side. TCP Vegas employs three key techniques to increase throughput and decrease losses:

- **New retransmission mechanism.** Vegas uses the system clock to provide more accurate RTT calculation[13]. Also the receiving and transmitting timestamps of the segments are recorded. When a duplicate acknowledgment is received, Vegas checks the time difference between current time and transmitted time for this segment. If this difference is greater than the timeout value, Vegas retransmits this segment without waiting for 3 duplicate acknowledgments

  Also if a non-duplicate acknowledgment is received immediately after a retransmission and the relevant timestamp is greater than the timeout value, the segment is retransmitted.

  The congestion window is only decreased when the retransmitted segment was previously sent *after* the last decrease of the congestion window. This avoids decreasing the congestion window more than once for losses that occurred during one RTT interval as is possible in TCP Reno.

---

[13] TCP Reno and Tahoe use a default timer granularity of 500 ms.

- **Congestion Avoidance Algorithm.** Instead of waiting for a congestion to react afterwards (as Reno does), Vegas tries to be proactive to really *avoid* congestion. TCP Vegas looks for changes in the sending rate to detect upcoming congestion. It compares the measured throughput with an expected throughput rate. A flattening of the measured throughput indicates an upcoming congestion where the algorithm can now proactively try to avoid the congestion.
- **Modified slow start mechanism.** Instead of increasing the congestion window in slow start exponentially *every* RTT, TCP Vegas increases the window only *every other* RTT. This means the slow start algorithm in Vegas is even slower than that of Reno or Tahoe.

This prevents, more efficiently, a possible large overshoot of the sending rate in the beginning of the connection, where no estimate of the bandwidth is available, and thus the slow start threshold value, ssthres, is not yet initialized. Thus the crossing point to the congestion avoidance phase is not yet defined.

2.3.2 User Datagram Protocol, UDP

UDP is a simple datagram-based transport layer protocol. It provides unreliable end-to-end connection. So the application itself needs to add reliability if it is needed. UDP is specified in [108]. The service is connectionless, i.e. no state information is maintained of the current connection at the sender or the receiver. UDP (as TCP) uses the functionality and services provided by the IP network layer protocol.

UDP is used in applications, where reliability is not the key. These are, for example, audio or video streams. Here, the loss of packets is not so severe as delay or jitter. As no retransmissions are done, the transmission time is only influenced by the underlying network and does not depend on protocol features.

The UDP header has a fixed size of 8 bytes. A sketch of the header is plotted in Fig. 2.13. The first two entries, *source port* and *destination port*

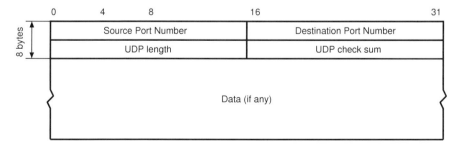

Fig. 2.13. Sketch of an UDP datagram including header and data portion

number are used to address the application in both endpoints of the connection. This is the same method as in TCP, however, the TCP and UDP port numbers are independent of each other.

The *UDP length* field is the length of the complete UDP packet (header inclusively data) in bytes. The minimum value for this field is 8 bytes. This information is redundant, because the IP header also provides the packet length. However, this can be seen as a strict separation between the two layers.

The *UDP check sum* covers the UDP header and data fields. The check sum algorithm is the same as in TCP. While TCP check sum is mandatory, UDP check sum is optional. But it is recommended to enable the check sum to avoid unnecessary packet processing in the upper protocol stack. As in TCP, the UDP check sum includes a pseudo-IP header including the IP source and destination address to check the entire UDP/IP check sum.

2.4 Application Layer

The application layer supports the user with common network applications or basic network procedures in order to implement distributed applications. Network applications are, for example, a file transfer standard or a virtual terminal. Commonly used application layer standards are *Telnet* for providing a virtual terminal, *FTP*, File Transfer Protocol for file transfer, *NFS*, Network File System for transparent access of files, *SMTP*, Simple Mail Transfer Protocol for mail delivery, *DNS*, Domain Name Service for centralized naming of Internet hosts and *WWW*, World Wide Web.

Internet today is often a synonym for access to the global distributed information in the net with WWW. The application layer protocol used for communication between WWW server and WWW client is the *Hypertext Transport Protocol, HTTP*. Because WWW Internet access is today by far the most important access method, we will restrict this introduction to HTTP.

2.4.1 Hypertext Transport Protocol, HTTP

The Hypertext Transport Protocol is an application-level protocol for distributed, collaborative, hyper media information systems. The current version of HTTP is HTTP 1.1, which is specified in [50].

HTTP is a generic protocol which allows basic hyper media access to resources available from diverse applications. Hence, HTTP simplifies the implementation of the user agent. It can be used for communication between user agents and proxies/gateways to other Internet protocols like SMTP, FTP.

HTTP allows an extensible set of methods and headers that indicate the purpose of a request. Messages are passed in a format similar to that used by Internet mail as defined by the Multipurpose Internet Mail Extensions (MIME) [74]. All HTTP requests are asynchronous. The communication usually is accomplished over TCP/IP connections, however HTTP can run on top of any *reliable* transport protocol.

Basic operation

The HTTP protocol is based on the request/response paradigm, i.e. a client/server structure. A client establishes a connection with a server and sends a request to the server in the form of a request method, including additional data like request modifiers, protocol version, client information and possible body content. The server responds with a status line. This line includes, besides other information, an error code, server information and possible body content.

In the simplest form the HTTP connection is opened between the client and the server directly. In more complicated situations, the client is connected to a proxy, gateway or tunnel entry point. A proxy is a forwarding agent which is receiving requests, then rewriting all or parts of them and finally it is forwarding the reformatted requests towards the server identified by the received destination address.

Features

In HTTP 1.0 almost always a separate connection (e.g. a TCP/IP connection) was created for every request/response exchange (e.g. for every image in a HTML page). HTTP 1.1 overcomes this problem by encouraging multiple requests on the same connection (*persistent connections*). Thus fewer connections are opened, CPU time is saved in routers and hosts, network congestion is reduced because TCP has sufficient time to react properly to congestion in the network.

The *pipelining* feature allows a client to send multiple requests to the same server without waiting for each response. Thus, again, the persistent TCP connection can be used much more efficiently. Additionally, HTTP does not need to implement flow control mechanisms itself but relies on the flow control of the TCP connection.

HTTP provides several optional *authentication* operations. These can be used by a server to request authentication of a client, and by a client to provide authentication information to the server. Detailed specification can be found in [73].

It is possible to *negotiate the content* type, either server or client initiated, to supply the user with the best available representation for a given request when there are multiple requests available.

Caching of responses in intermediate systems (e.g. proxies) is supported by HTTP by special protocol elements.

2.5 Wireless Application Protocol, WAP

The Wireless Application Protocol (WAP) is a set of protocols, application environments and content formats that builds upon current and future bearer services. WAP is (besides any adaptation) independent of the bearer service used. WAP should enable mobile terminals to access the Internet with usage of standardized services, protocols and formats. The WAP protocols and data formats are adjusted for the special requirements in mobile radio and to the client server architecture of the Internet.

The Wireless Application Protocol is specified by the WAP-Forum. The WAP-Forum was founded in December 1997 by Nokia, Ericsson, Motorola and Phone.com (formerly Unwired Planet). Today, the WAP-Forum has more than 100 companies as members and associate members. Members are manufacturers from network infrastructure and end-user terminals but also software manufacturers and mobile radio network operators, among them are all the notable companies[14].

Version 1.0 of WAP was published by the WAP-Forum in April 1998. In May 1999 Version 1.1 was presented and in November 1999 Version 1.2. The specification is public and can be found on the Internet (http://www.wapforum.org).

2.5.1 Architecture Overview

The WAP architecture is very similar to the WWW architecture. The WWW model specifies a client-server architecture. The client requests information from a server that answers with a response (see also Sect. 2.4.1). Normally no other entity is involved in the communication path[15].

The WAP model specifies also a client-server architecture with an additional gateway in the communication path between client and server (Fig. 2.14). Between the client and the gateway *encoded* communication by means of the WAP protocol stack is accomplished. The gateway translates requests from the WAP stack to the WWW protocol stack and sends it further on to the origin server. On the way back from the origin server to the client, the gateway also encodes the content into compact format suitable for communication over low-capacity (wireless) links. Gateway and origin server may reside physically in the same entity (i.e. computer).

---

[14] The actual list can be found under http://www.wapforum.org.

[15] However, three classes of servers are defined in the WWW. A server could be an origin server. This is where the requested resource (content) resides or is to be created. Second class is the proxy server: An intermediate program that acts as both server and client. It captures the request from clients and processes it either locally or passes it to other servers. Third class is a gateway: It acts as an intermediary for some other server. Unlike a proxy, a gateway behaves like the original server, i.e. the client may not be aware that it is communicating with a gateway and not with the requested server itself.

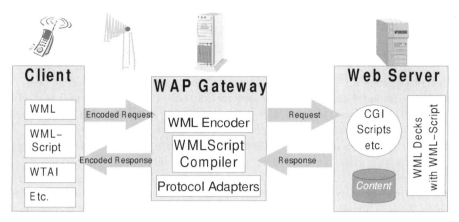

Fig. 2.14. WAP programming model

WAP defines a set of standard components including a naming model, content typing, standard content formats and standard communication protocols. WAP uses existing standards wherever they are useful. A *micro browser* in the wireless terminal co-ordinates the user interface and is analogous to a standard web browser.

WAP can provide end-to-end security between WAP protocol endpoints. If a client and origin server desire end-to-end security, they must communicate directly using the WAP protocols. Alternatively, the gateway server could be trusted by the origin server or physically located at the same place as the origin server.

The WAP protocol stack is drawn in Fig. 2.15. We will provide a short description of each layer and the according services and protocols in the next sections.

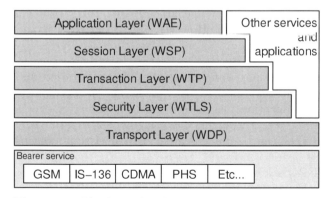

Fig. 2.15. WAP layered architecture and protocol stack

2.5.2 Transport Layer

The transport layer protocol in the WAP framework is the *Wireless Datagram Protocol*, WDP. WDP is defined in [137]. It operates above a data-capable bearer service. WAP is designed to operate over a variety of different bearer services, including short-message service, circuit-switched data and packet-switched data. Every bearer offers different quality of service. WAP protocols are designed to compensate or tolerate these varying levels of service.

WDP provides a general datagram service to upper-layer protocols and communicates transparently over one of the available bearer services. As WDP provides a common interface to the upper layer it hides the differences of the underlying bearer service by adaptation to the different bearers (Fig. 2.16). Currently there exists adaptation for all important mobile ra-

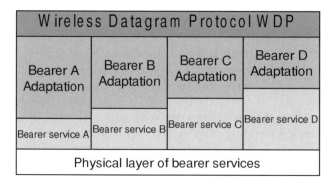

Fig. 2.16. Adaptation of the WAP protocol stack to the bearer service by the WDP protocol

dio standards, like GSM-CSD, GSM-SMS, GSM-GPRS, GSM-USSD[16] , IS-136[17], PHS[18]. As every bearer has different characteristics, WDP protocol performance may vary over each bearer. However, the WDP services and service-primitives will remain the same.

WDP service is an *unreliable* service. Besides the datagram service it offers application addressing by port numbers, optional segmentation and reassembly of large datagrams and optional error detection. WDP supports several simultaneous communication instances from an upper layer over a single underlying bearer service. The higher layer instance is identified by the port number. Also, WDP can be implemented to support simultaneous connections over multiple bearer services.

If an error in the WDP stack occurs, the *Wireless Control Message Protocol*, WCMP, [136], should be used. WCMP provides error handling mecha-

---

[16] Unstructured Supplementary Service Data.
[17] Interim Standard 136. American mobile radio standard.
[18] Personal Handyphone System. Japanese mobile radio standard.

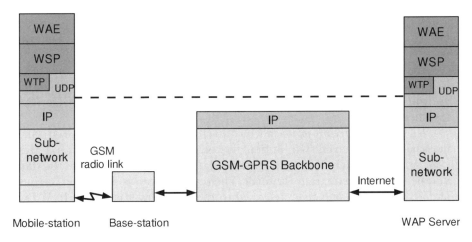

Fig. 2.17. WAP protocol profile for GSM-GPRS bearer service. As GPRS uses IP as network layer protocol, WDP is identical to the User Datagram Protocol, UDP

nisms for errors like there is no application listening to the specified destination port, or the buffer space in the receiver is too small for a large message.

If the underlying bearer service uses the Internet Protocol (IP), the WDP protocol *must* be the User Datagram Protocol (UDP) and WCMP is *identical* to the Internet Control Message Protocol (ICMP)! For bearer services not supporting IP the WDP protocol defined in [137] must be used. An important case where the bearer service uses IP is GSM-GPRS. The protocol stack profile is illustrated in Fig. 2.17. In this configuration UDP (which operates as WDP) works above the Internet Protocol, IP. The Wireless Transaction Protocol, WTP, and optionally also higher layers use the UDP datagram service as transport layer. IP is the network layer protocol used in GSM-GPRS (see Sect. 3.2 for further details on the GSM-GPRS system).

2.5.3 Security Layer

The security layer protocol in the WAP framework is called the *Wireless Transport Layer Security*, WTLS. The WTLS layer operates above the transport layer. WTLS provides a secure transport service which can be used by upper-layer protocols (e.g. Wireless Transaction Protocol, WTP, or Wireless Session Protocol, WSP). The security layer in WAP is *optional*. The WTLS layer provides privacy, data integrity and authentication between the two communication endpoints. WTLS is derived from the Transport Layer Security protocol, TLS, [46]. Thus its functionality is similar to TLS 1.0. But WTLS is optimized for low bandwidth and high-latency bearer networks, so some new features are added like datagram support, optimized handshake and dynamic key refreshing.

WTLS provides an independent connection management that allows a client to connect with a server and to negotiate protocol options. The connection establishment consists of several steps. The negotiation may include security parameter, key exchange and authentication. When the connection is functional, it is in a specific connection state. There is a compression algorithm, encryption algorithm and Message Authentication Code algorithm specified.

WTLS uses a *record protocol* which itself has a layered structure. It takes massages to be transmitted, optionally compresses the data, applies a Message Authentication Code, encrypts the data and finally transmits the data via the Wireless Datagram Protocol. The received data is decrypted, verified and decompressed and then delivered to the upper-layer protocol. In WTLS four record protocol clients are described: The *change cipher spec protocol*, the *handshake protocol*, the *alert protocol*, and the *application data protocol* (see Fig. 2.18). In essence, the change cipher spec protocol and the alert pro-

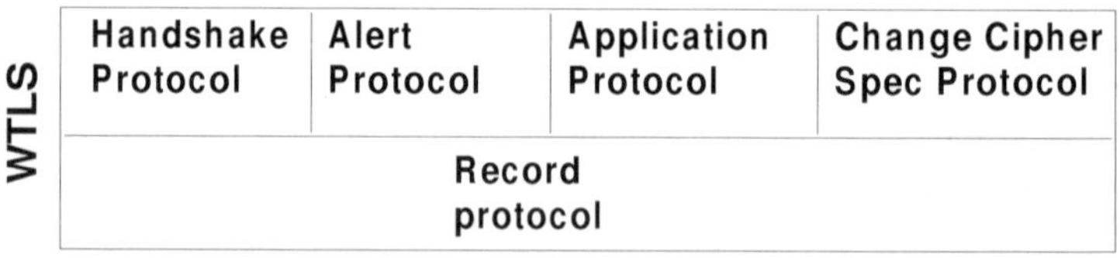

	Handshake Protocol	Alert Protocol	Application Protocol	Change Cipher Spec Protocol
WTLS	Record protocol			

Fig. 2.18. WTLS record protocol. It consists of four different protocols used for connection establishment, exception handling, data transmission and security specification change (from left to right)

tocol are parts of the handshake protocol. The handshake protocol involves the following steps [145]:

- Exchange 'hello' messages to agree on algorithms and exchange random values.
- Exchange the necessary cryptographic parameters to allow the client and server to agree on a pre-master secret.
- Exchange certificates and cryptographic information to allow the client and server to authenticate themselves.
- Generate a master secret from the pre-master secret and exchanged random values.
- Provide security parameters to the record layer.
- Allow the client and server to verify that their peer has calculated the same security parameters and that the handshake occurred without tampering by an attacker.

Because WTLS operates over the unreliable datagram service WDP, special precautions are used: WTLS requires that the handshake messages in one direction are concatenated in a single transport service data unit. The client must retransmit the handshake message if necessary. The server must appropriately respond the retransmitted messages from the client.

A number of different standard Internet cryptographic algorithms can be used for key exchange (Symmetric, RSA, Diffie–Hellman, Elliptic Curve Diffie–Hellman, ...), bulk encryption (Data Encryption Standard DES, ...) and message authentication (MD5,...).

Additionally, a Wireless Identity Module, WIM, [138], is available to achieve an even better security standard. An example for WIM could be a smart card. It can be the Subscriber Identity Module or a separate external smart card. This identity module could hold long-living master keys and other security-relevant information like information for user authentication.

2.5.4 Transaction Layer

A transaction is defined as a request/response service necessary especially for interactive applications. The *Wireless Transaction Protocol*, WTP, is the transaction layer protocol in the WAP architecture. It is specified in [144]. WTP is designed to provide a *reliable* transaction service while balancing the amount of reliability required for the application with the cost of delivering the reliability. WTP is message oriented and designed for interactive applications, such as 'browsing' the Internet. WTP operates above the unreliable datagram service WDP or optionally above the transport security layer.

Three different classes of transaction services are defined:

- Class 0: Unreliable invoke message with *no* result message.
- Class 1: Reliable invoke message with *no* result message.
- Class 2: Reliable invoke message with exactly one reliable result message.

Reliability is achieved by usage of unique transaction identifiers, acknowledgments, duplicate removal and retransmissions. The unit for transaction identifiers is a 'transaction'. For class 1 and 2 transactions, every message must be acknowledged not only from the protocol but optionally also from the service user (i.e. the upper-layer protocol that uses the transaction service). This improves user-to-user reliability. If an error occurs (either the message is received corrupt or the message is lost, which is trapped by a retransmission timer at the sender), the message is resent as long as the retransmission timer is below a certain threshold. However, WTP has no error detection but relies on lower-level protocols like WDP.

In WTP there is no explicit connection setup or tear-down phase intended. Thus protocol overhead for short transactions is significantly improved.

Also, asynchronous transactions are supported. The responder may signalize the initiator of the transaction that it cannot serve the request immediately ('hold on' acknowledgment). It is thus allowed to wait until the requested response is available and then sends its response. The initiator does not have to make a second request.

Optionally, segmentation and reassembly of messages is supported. If the message length exceeds the maximum transmission unit, MTU, of the current bearer service, the message can be segmented into several packets. If a large number of packets is created the packets are sent in groups. Furthermore, not every packet must be acknowledged but only every packet group. For sending of packet groups a Stop-And-Wait protocol is used. A new packet group is only allowed to be sent when the previous group is fully acknowledged. A small amount of flow control can be exercised by the sender by changing the size of the packet groups.

2.5.5 Session Layer

The *Wireless Session Protocol* is the session layer protocol in the WAP framework. It is specified in [140]. WSP uses both the datagram service of WDP and the transaction service of WTP. It provides the application layer the means for organized exchange of content between client and server applications. WSP maintains long-lived sessions. Specifically the following functions are supported: Connection establishment and resuming, protocol negotiation capability, and compact encoding for content exchange.

Two different services are currently defined in WSP:

- Connection-mode session service over a transaction service. This provides a reliable connection but requires explicit session establishment.
- Connectionless service over a datagram transport service. This is an unreliable connection only and can be used without having to establish a session.

WSP provides HTTP/1.1 functionality, i.e. extensible request-reply methods, composite objects and content type negotiation. All HTTP/1.1 methods are supported. However, the protocol information is binary encoded. Also WSP has a more efficient capability negotiation than HTTP.

A session may be open if the underlying transport service has been removed, as well, i.e. the connection has been closed. Then the session is in a suspended state. But all relevant session parameters are held constant during suspended state. When the connection is re-established again, a lightweight connection re-establishment protocol allows the session resume without the total overhead of a complete new session establishment. A session may be resumed also over a different bearer network.

In addition to the HTTP request response mechanism, WSP provides a *push* mechanism for data transfer. This allows the server to push data to the client at any time during the session. Confirmed and non-confirmed data push methods are defined.

Optionally asynchronous requests are supported. The client may submit multiple requests without waiting for immediate response. Thus latency is improved because the server can send the requested information immediately as it becomes available. Also bandwidth usage is optimized because multiple requests can be concatenated into fewer messages.

2.5.6 Application Layer

The *Wireless Application Environment*, WAE, specifies an application framework for wireless devices. It uses and extends other WAP technologies, (WSP, WTP, ...) as well as existing Internet standards such as XML (Extensible Markup Language), URLs (Uniform Resource Locator), scripting language and various content formats. The general WAE model is shown in Fig. 2.19 The WAE is divided into two logical layers:

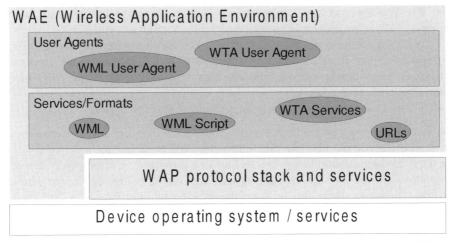

Fig. 2.19. Model of the Wireless Application Environment. It consists of User Agents and Services and Formats and operates above the WAP protocol stack and the operating system of the device

- *User agents* are entities that interpret the services and formats. The WML user agent is a fundamental user agent within the WAE. It can be considered as a micro browser in the WAP architecture. But WAP does not specify the implementation of user agents nor are there any restrictions in specifying new user agents. In particular, the Wireless Telephony Application extension (see below in this section) specifies a user agent which allows access and interaction with mobile telephony features.
  Also, features and capabilities of user agents are left to the implementors. WAE only defines basic functionality (i.e. services and formats) to ensure compatibility between implementations.
- *Services and formats* contain common elements and formats that are accessed and interpreted by user agents. Included are the Wireless Markup Language, WML, WML script, image formats, calendar formats or the like.

Analogous to the logical interpretation in a computer system, the user agents can be considered as the 'programs' that interpret the 'data' (services and formats) in a specified manner.

In the following we will provide a very brief description of the basic WAE services and formats.

Wireless Markup Language, WML

WML is a tag-based document description language that has been derived from HTML [41] and the Handheld Device Markup Language, HDML [107], which was developed by Unwired Planet (now Phone.com), a founder of the WAP-Forum. WML has been developed for optimized presentation of content and for user interaction on devices with limited capabilities. It is specified in [139].

The Wireless Markup Language is based on the Extensible Markup Language, XML, [146], i.e. it is specified as XML document type.

WML has a number of features, which are: Support for text and images, support for user input, navigation and history stack, international support, Man-Machine-Interface independence, narrow-band optimization and state and context management.

The functions of WML can be summarized into four areas:

- Text and image presentation and layout. This also includes a number of formating commands, like highlighting a text, etc.
- Logical organization in the Deck and Card metaphor. Cards specify one or more units of user interaction. A user can navigate through cards where links between cards can be used for moving to another card. Decks are groups of cards. A WML deck can be compared with a HTML page. It is also the unit of transmission. A deck can be addressed with a Uniform Resource Locator, URL.
- Navigation between cards is supported through links. Event handling mechanisms can be used for navigation or execution of scripts.
- String variables allow run-time substitution of strings. This allows for improved efficiency on network demands.

WMLScript

WMLScript is a simple scripting language designed to overcome the limitations of WML to provide user interactivity. WML pages are static and no reaction to user input is possible. WMLScript extends WML with a programmable part. It is specified in [139].

WMLScript introduces improved user-interface handling, e.g. error messages, dialogs like alerts or confirmation can be produced locally. The device can react on user interaction. Thus it reduces the network load by moving more intelligence into the device.

Wireless Telephony Application, WTA

The Wireless Telephony Application, WTA, is an extension to the Wireless

Application Environment, which makes it possible to use telephony applications from within the WAP framework. WTA is defined in [143].

With WTA it is possible to access mobile network telephony features, such as setting up a call, from within WAP applications, i.e. from a WML Deck or controlled by a WMLScript.

WTA introduces a new user agent within the WAE with capabilities for interfacing to mobile network features, like call submission or the like. The WTA extension to the general WAP architecture is illustrated in Fig. 2.20. WTA elements include also a WTA server, WTA Interface (WTAI) and a

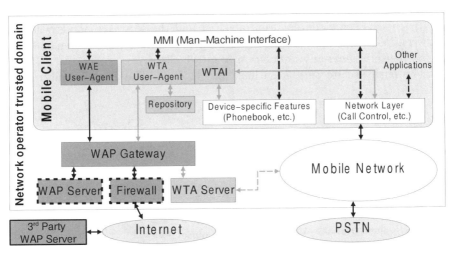

Fig. 2.20. Model of the WAP architecture with Wireless Telephony Application extension. These are the WTA user agent, WTA server, WTA interface (WTAI) and Repository

Repository.

The WTA user agent gets its content only from the WTA server or from the Repository. The Repository is a memory in the device to enable fast access to immediately needed content. This is necessary in WTA to provide real-time interaction with the mobile network. WTA can access the mobile network features and device-specific features, like phone book, through the WTA Interface WTAI. The WTAI is network specific, e.g. the interface for GSM is specified in [142] whereas the common part for all network types is defined in [141]. The interface between the WTA server and the mobile network is beyond the scope of the WTA specification.

As considerable knowledge of the mobile network is required for WTA, most of the features are only accessible by the network operator. Only a limited, basic set of WTA functions (e.g. initiating a phone call) is available to third-party authors. But *only* trusted service providers can use this features which means that WTA services can only be loaded from trusted origin servers. This implies a separate WTA security model that the mobile network operator has to specify.

WTA services can be initiated in three different ways: An URL is accessed (either via the network or in the repository) which requests a WTA service, a WTA service indication is sent from the network (i.e. the network initiates a WTA service, also called 'push' service) or a mobile network event occurs (e.g. a new call is arriving) and this event is bound in WTA to specific action.

2.5.7 WAP Push Architecture

In contrast to the common client-server architecture in WWW and WAP (a client requests some content from a server and the server responds with an answer), WAP also specifies a pure 'Push' service. Here, the server initiates the response *without* any interaction from the client.

The push architecture is outlined in Fig. 2.21. The Push Initiator con-

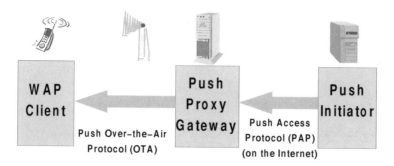

Fig. 2.21. Model of the WAP push architecture

tacts the *Push Proxy Gateway* using the *Push Access Protocol* and transmits content to this gateway. The proxy translates the content and sends it to the WAP client via the *Push Over-the-Air Protocol*. The push services and involved entities and protocols are specified in several documents ([130–135]).

The **Push Access Protocol, PAP**, is used by the Push Initiator to submit a push message to the proxy. Also, other operations like notification of the result, cancellation and querying the status of an already submitted push message and querying the capabilities of a client are supported. The Hypertext Transport Protocol, HTTP, has been chosen initially as carrier protocol, although this protocol can be tunneled through any other common Internet protocol. PAP messages are XML-style entities which can also be grouped in a multi-part document.

The **Push Proxy Gateway, PPG** is the entry point to the mobile network. The owner of the gateway decides the access restrictions for push services to WAP clients in the mobile network. The PPG translates the messages coming from the initiator into the Push Over-The-Air Protocol to improve bandwidth efficiency. PPG functionality may be built into the normal WAP gateway to reduce the necessary resources and to gain in efficiency with using shared session over the air interface.

The **Push Over-the-Air Protocol** operates on top of the Wireless Session Protocol (WSP), i.e. it uses WSP sessions for content delivery. The Push OTA Protocol is responsible for push content delivery to WAP clients, application addressing, exchange of push control messages over the air interface, bearer selection and authentication of a push initiator. Push transfers may be performed in either connection-oriented or connectionless manner. For a connection-oriented mode a WSP session must be active before push content is delivered. If no session is active the OTA server may initiate a session on the WAP client with the *Session Initiation Application*. However, the WAP client may verify if this server is allowed to open a new session.

The Push framework provides means for **service loading**. A push initiator may send a push message to a WAP client with a *Service Loading*, SL, message. The SL message includes a URI (Uniform Resource Identifier) which is now automatically requested from the origin server (or the cache in the client) by means of the conventional WAP client/server concept, without any user interaction. When the URI is received by the client the service starts executing. Thus a mobile network operator can push new services (e.g. reload prepaid cards) to the WAP client. Also **Service Indication** is specified in the push framework. It enables push initiators to notify clients about new events, e.g. news headlines, e-mails or the like.

2.5.8 Comparison WAP vs. Internet Protocol Stack

In this section we compare some protocols used in the Internet with those defined in WAP. Namely, we compare TCP and UDP with WTP and WDP. It is not always possible to associate an Internet protocol completely and without ambiguity to a specific WAP layer or WAP layer protocol. Thus the following comparison may be not the only possible way to do it. We provide two different possibilities.

The first comparison is shown in Fig. 2.22. The three lowest layers, physical, datalink and network layer are not specified within WAP. Thus our comparison starts at the transport layer.

In WAP, the Wireless Datagram Protocol, WDP, is identical to UDP if the used bearer service supports IP (see Sect. 2.5.2). In the case of other bearer services, the differences between WDP and UDP are mainly the maximum possible packet length and the packet header length (i.e. the protocol overhead). Additionally, WDP provides optional segmentation of packets and also optional error detection, both functionalities not provided by UDP. Nevertheless, UDP can be seen as analogue to the WDP protocol.

A more differentiated sight is necessary for TCP. As TCP provides a reliable transport service, and the lowest reliable service in WAP is given by WTP, it is reasonable to compare them: The main differences between TCP and WTP are:

- TCP is byte-stream-oriented while WTP has 'messages' as the basic transmission unit.
- TCP utilizes a 3-way handshake when setting up a connection. WTA in contrast has *no* explicit connection set up or tear-down phase. Thus TCP is more suitable for long-lasting connections, while WTP is optimized for short connections with relatively small amount of data.
- For providing reliability, both protocols use retransmission of packets/-messages in the case of errors until the packet/message is acknowledged. However, TCP has its own error detection whereas WTP relies on the error detection of the lower layer. The retransmission timer (used in the sender to retransmit packets when the acknowledgment is outstanding longer as this timer) as adapted to the actual transmission characteristics (RTT) only in TCP, no such efforts have been made for WTP.
- To achieve reliability against duplicates and out-of-order delivered packets, both protocols use numbering of packets. TCP numbers each byte and WTP uses a unique 'Transaction Identifier' for each message sent.
- Flow control is very sophisticated as implemented in TCP (see Sect. 2.3.1). A sliding window mechanism with dynamical window adjustment is used. Such functionality is very restricted in WTP. Flow control is only possible when segmentation and reassembly is switched on (which is optional). The PDU is segmented and the resulting small packets are grouped together.

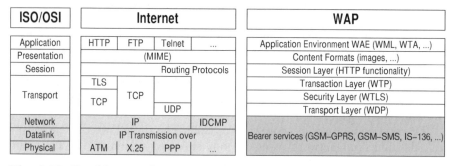

Fig. 2.22. Possible way of comparing the Internet protocol stack with the WAP protocol stack. On the left side the ISO/OSI layered communication model is also drawn for clarification

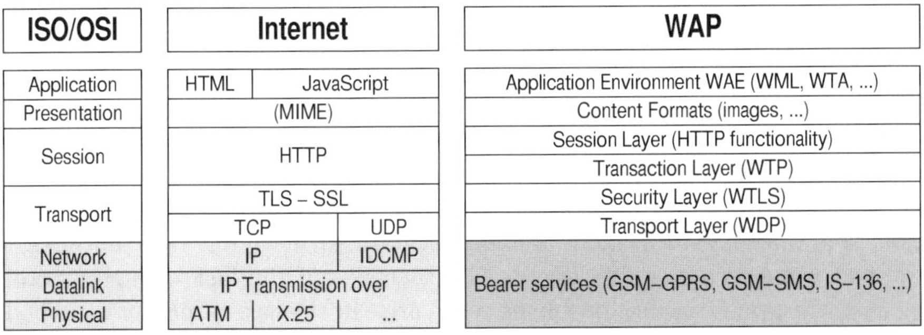

ISO/OSI	Internet		WAP
Application	HTML	JavaScript	Application Environment WAE (WML, WTA, ...)
Presentation	(MIME)		Content Formats (images, ...)
Session	HTTP		Session Layer (HTTP functionality)
			Transaction Layer (WTP)
Transport	TLS – SSL		Security Layer (WTLS)
	TCP	UDP	Transport Layer (WDP)
Network	IP	IDCMP	
Datalink	IP Transmission over		Bearer services (GSM–GPRS, GSM–SMS, IS–136, ...)
Physical	ATM	X.25 ...	

Fig. 2.23. Another possible way of comparison of the Internet protocol stack with the WAP protocol stack. On the left side the ISO/OSI layered communication model is plotted for comparison. The difference from the first model (Fig. 2.22) is that HTTP is seen as a session layer protocol, thus TCP is assigned to the lower transport layer

Each group is sent by using a stop-and-wait protocol[19]. The only means for flow control is to change the size of the groups.

Thus again, WTP seems to be inadequate for transferring large amounts of data.

The security layer in the Internet (Transport Layer Security, TLS) normally operates above the TCP protocol. WTLS, the WAP security layer protocol is located between the transport layer and the transaction layer. Nevertheless, both protocols provide nearly the same functionality.

The session layer functions in WAP provide the same elements as HTTP in the Internet world. However, HTTP is intended to be an 'application layer protocol' [50]. Thus the application layer functions of the Internet have been moved to the session layer in WAP.

The content formats in WAP (images, phone book, calendar format) can be assigned to the presentation layer, as it provides a common data representation between two endpoints.

WAE, the application environment in WAP is formally associated to the application layer, even though no real protocols are defined therein. Nevertheless, many functions or languages defined in WAE (like WML) have equivalents in the Internet world, although those are usually not considered to be contained in the OSI protocol stack.

A second possible comparison is shown in Fig. 2.23 [49]. The differences from the first point of view are the following ones:

- HTTP is now seen as a session layer protocol.
- TCP and UDP are assigned to the WDP layer only. But TCP can not be compared to WDP reasonably.

---

[19] A stop-and-wait protocol sends the next packet only when the previous packet has been positively acknowledged.

- HTML and JavaScript are considered as 'application protocol'.

Conclusion

Based on the comparison above we draw the following conclusions:

The Wireless Datagram Protocol, WDP, is very similar to the User Datagram Protocol, UDP, thus the performance should be almost identical.

WTP, the transaction protocol in WAP, is based on message transactions and optimized for small amounts of data. If more data is transferred, the performance is possibly lower. Also, because of the lack of flow-control mechanisms, performance problems could arise in the case of packet-switched mobile networks, e.g. GPRS. TCP reacts upon congestion of a network accordingly (setting back its transfer rate) but WTP has no such mechanisms. Thus if many transfers are made simultaneously over the base station (or in GPRS the same Serving GPRS support node, SGSN) congestion could occur at this link if it is not dimensioned well. If this is the case, WTP will not detect this as a problem *if* it sends transactions with very large message size.

Thus, for future high-performance mobile networks the performance of WAP could be a possible bottleneck.

3. Data Transmission in GSM

In this chapter we will describe how data transmission is performed within the GSM system. First we provide the basics of circuit- and packet-switched data modes (Sects. 3.1 and 3.2). Also the protocols used are briefly explained. To determine performance of the GSM link, both uncoded and with channel coding, we made link level simulations on bit level (Sect. 3.3). Therein we used a realistic channel model (COST 207 [96]). However, bit-level simulations are very time consuming and not suited as an underlying module for simulations of higher-layer protocols (which we present in Chap. 5). Thus we developed a block error model from the link level simulations that is used subsequently as replacement for the bit-level simulations (Sect. 3.4). At the end of this chapter we will give a summary.

The Global System for Mobile Communication (GSM) is by far the most important digital mobile radio standard for public usage today. GSM has been standardized by the European Telecommunications Standard Institute (ETSI). The entire GSM standard consists of numerous parts, known as GSM Recommendations. As GSM is very complex, we can provide only a summary of those parts that are relevant for this book. This covers mainly functions and entities necessary for data transmission, like protocol architectures and rate adaptation. Not included either are other vital functions like the modulation, synchronization, signalling, radio link control, authentication, mobility management. For more detailed information about GSM we refer to the GSM Recommendations[1] as well as some books, [47, 101, 129].

GSM can be viewed as a hierarchical network which consists of three major subsystems, the Radio Subsystem, the Network and Switching Subsystem and the Operation and Maintenance Subsystem ([62]). A (simplified) GSM network is outlined in Fig. 3.1.

The radio interface ([68]) of GSM (Um-Interface) uses Time and Frequency Division Multiplex (TDMA and FDMA) for multiple access of mobile stations. One physical channel is characterized by the carrier frequency and a *timeslot* in a *frame*, which is repeated every 4.613 ms. Every frame is divided into 8 timeslots with a duration of 577 µs, where different kind of *bursts* are transmitted. Prior to GSM Phase 2+, only a single timeslot per frame could have been used for one traffic channel. The raw radio bit rate per timeslot,

---

[1] They can be found on the Internet, http://www.etsi.org/.

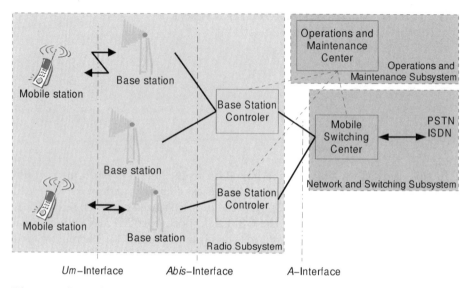

Fig. 3.1. Simplified overview of a GSM network

when using a normal burst (114 raw user data bits), is 24.7 kbit/s. With introduction of Phase 2+ also more timeslots per frame can be allocated, thus allowing higher radio bit rates per traffic channel.

The transmission of user data (Traffic Channels, TCH) and signalling information is organized in logical channels. For the mapping of logical channels to a physical channel a hierarchical multiframe structure is used. Because synchronization, idle, and other control channels are multiplexed together with the traffic channels in the multiframe, the final available raw data rate for traffic channels is 22.8 kbit/s.

Data transmission services, in general, can be classified in *circuit-switched data* and *packet-switched data* services, according to their switching paradigm. In circuit-switched data services (CSD) a 'virtual line' is allocated between the two transmission endpoints, and this line remains open during the entire transfer. Every data portion sent is routed along the same way. CSD services could be symmetric or asymmetric, which means the bandwidth is either the same in forward and backward directions or not. In contrast to CSD, packet-switched data services have *no* fixed connection, and every packet is transmitted independently of each other. Packet switched services are more naturally suited to bursty traffic, as no data need to be transferred when there is no data to send.

In GSM initially only circuit-switched services have been standardized. We will describe these services in the following section. In the GSM packet-switching extension, *General Packet Radio Service* (GPRS), introduced with GSM Phase 2+, also a number of packet-switched services are defined. GPRS, however, is a major upgrade of the GSM overall structure and the core network. We will present GPRS briefly in Sect. 3.2.

3.1 Circuit-Switched Data

3.1.1 Basics

In GSM Phase 2 different traffic channels have been standardized, from 2.4 kbit/s up to 9.6 kbit/s user data rate [58]. There exist full-rate and half-rate traffic channels. Full-rate channels (TCH/F) require a full raw radio interface rate of a single timeslot, half-rate channels only every other timeslot. Thus two half-rate channels require the same resources as one full-rate channel. In GSM Phase 2+ additionally the TCH/F14.4 traffic channel with 14.4 kbit/s and also, optional multislot operation have been standardized. Circuit-switched multislot operation is called *High-Speed Circuit-Switched Data* (HSCSD). The main aspects of HSCSD are summarized in Sect. 3.1.4. Due to the importance of the 9.6 kbit/s full rate (TCH/F9.6) and the 14.4 kbit/s (TCH/F14.4) traffic channel we will restrict the following explanation to them.

Traffic channel TCH/F9.6
Due to the multiframe structure, i.e. signalling, the total coded radio interface rate of a TCH/F9.6 channel aggregates to 12 kbit/s, i.e. 2.4 kbit/s are used for signalling and 9.6 kbit/s are available as user data rate. The difference between the radio interface rate (12 kbit/s) and the raw data rate (22.8 kbit/s) is used for Forward Error Correction (FEC). FEC consists in essence of a punctured convolutional code with code rate 1/2 and an interleaver with an interleaving depth of 19 bursts. See Fig 3.2 for illustration.

Traffic channel TCH/F14.4
With a reduced number of signalling and status bits a user data rate of 14.4 kbit/s at a radio interface rate of 14.5 kbit/s has been achieved. Together with a not so powerful convolutional code (same generator polynomials then for TCH/F9.6 but puncturing is increased) this higher radio interface rate is mapped to a full rate channel (Fig. 3.2). However, this higher net bit rate increases the required minimum radio transmission quality (C/I_{min}, carrier to interference ratio) for reliable transmission. See Sect. 3.3.3 for carrier-to-interference ratio (C/I) requirements for a specific BER.

Every traffic channel is available in two different modes:

- **Transparent mode.** In transparent mode just layer 1 functionality is used for data transmission. The only way for error correction is FEC. The user bit rate and delay is constant but Bit Error Ratio (BER) depends on the link quality. The protocol stack for transparent, asynchronous data transmission is shown in Fig. 3.3. At the mobile station (MS) the asyn-

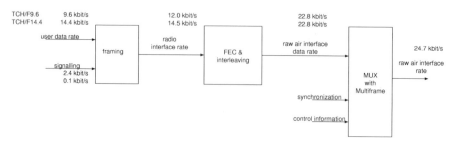

Fig. 3.2. Relation between net user data rates, radio interface rate and raw air interface rate for TCH/F9.6 and TCH/F14.4 traffic channels

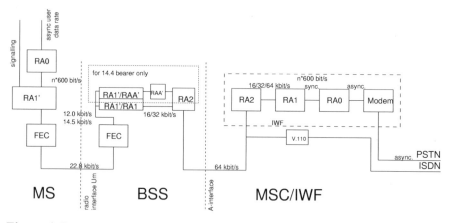

Fig. 3.3. Transparent, asynchronous data transmission in GSM [129]

chronous data stream is converted into a synchronous data stream of a rate n×600 byte/s by the *Rate Adaptation* (RA) function RA0. The rate adaptation functions are defined in [55] and [57]. This is further converted into the radio interface rate of 12 kbit/s (for TCH/F9.6) or 14.5 kbit/s (for TCH/F14.4) by function RA1'. RA1' also includes the signalling information of the asynchronous user data stream. The forward error correction (FEC) expands this to the raw radio interface rate of 22.8 kbit/s.

In the base station subsystem (BSS) the radio interface rate is converted to a fixed rate of 64 kbit/s at the *A*-interface, which is the only rate a MSC can process. Different RA functions are applied for TCH/F9.6 and TCH/F14.4. For the TCH/F9.6 bearer service the radio interface rate is converted into an intermediate rate of 16 or 32 kbit/s and further on to 64 kbit/s. In TCH/F14.4 service two intermediate rates are specified before going to fixed 64 kbit/s ([47, 129]).

At the *Mobile-services Switching Center* (MSC) an *Interworking Function* (IWF) reconverts the 64 kbit/s synchronous data stream into an asynchronous stream. The modem modulates this stream for transmission over the Public Switched Telephone Network (PSTN). For further details regarding the IWF we refer to Sect. 3.1.3. In the case of standard ISDN (Integrated Services Data Network) only a rate adaptation according ITU V.110 is necessary for binding to the ISDN network.

- **Non-transparent mode.** In the non-transparent data transmission mode in GSM a layer 2 protocol, the *Radio Link Protocol* (RLP), is responsible for error correction. RLP is in addition to FEC. RLP is a selective repeat ARQ protocol. It is described in more detail in Sect. 3.1.2. Figure 3.4 shows the protocol stack for non-transparent asynchronous data transmission.

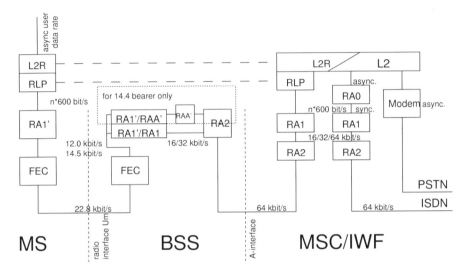

Fig. 3.4. Non-transparent, asynchronous data transmission in GSM [129]

The asynchronous user data stream is transferred to the *Layer 2 Relay* (L2R) protocol. L2R is responsible for an asynchronous, character-oriented connection between the L2R endpoints at the mobile station and the MSC. The L2R Protocol Data Units (PDUs) have fixed length corresponding to the RLP frame information field. L2R uses the reliable transmission service of RLP. The L2R Character Oriented Protocol also transmits the status information of the asynchronous lines at the endpoints.

Between rate adaptation function RA1' at the mobile station and RA1 at the MSC/IFW the transmission plane and rate adaptation functions are the same as in transparent mode. After RA1 at the MSC/IWF the RLP and L2R protocols are terminated. Thus a reliable connection exists between MS and MSC already at layer 2. L2R is the intermediate entity for communicating with PSTN (via a modem) or with ISDN (via rate adaptation and synchronization functions RA0, RA1 and RA2).

Data connection to the standard ISDN network (i.e. without any modems) is also called *Unrestricted Digital* (UDI) connection. For connections via the PSTN a modem at the fixed end point is applied just as for traditional data connections within a PSTN. This uses the 3.1 kHz services of the PSTN ('3.1 kHz Audio').

3.1.2 Radio Link Protocol (RLP)

The Radio Link Protocol ([64]) provides a reliable connection between mobile station and MSC/IWF in the case of non-transparent data transmission modes (see Fig. 3.4). Retransmissions of frames with errors are applied for reliable data transmission. Thus the user data rate is *not* constant any longer, but the BER becomes very low (of the order of 10^{-9}). RLP is a layer 2 protocol, i.e. it provides a data link service. It is based mainly on the ISO High level Data Link Control (HDLC) protocol and has been adopted to the special needs for radio transmission.

Three versions of RLP are defined:

- RLP version 0: single-link basic version
- RLP version 1: single-link extended version (e.g. added data compression feature)
- RLP version 2: multi-link version

Single-link versions use a single physical link, the multi-link version supports transmission over multiple physical links in parallel. This feature is used in HSCSD (Sect. 3.1.4). All RLP versions are backwards compatible. RLP is fully duplex, i.e. simultaneous transmission in both directions is possible.

As RLP operates on top of FEC it assumes a channel with a BER below approximately 10^{-3}. Together with an RLP block length of 240 bits (TCH/F9.6) or 576 bits (TCH/F14.4) a block error rate of less than 10^{-1} is assumed for RLP to perform adequately. For frames sent on different physical channels in multi-link operation, the time difference between two frames must not exceed a specific parameter $T4$. This is recommended to be set to 30 ms (TCH/F9.6) or 50 ms (TCH/F14.4).

The basic RLP block structure is illustrated in Fig. 3.5. A frame consists

	Header	Information	FCS
240 bit frame size			
Vers.0,1; Vers 2 (U frames only)	16 bit	200 bit	24 bit
Vers 2 (S and I+S frames only)	24 bit	192 bit	24 bit
576 bit frame size			
Vers.0,1; Vers 2 (U frames only)	16 bit	536 bit	24 bit
Vers 2 (S and I+S frames only)	24 bit	528 bit	24 bit

Fig. 3.5. Basic RLP frame structure [64]

of a header, an information field and a Frame Check Sequence (FCS) for error detection. The length of the various fields depend on the overall frame size (i.e. TCH/F9.6 or TCH/F14.4) and the frame type. Three different frame types are defined: U-frames (unnumbered) are used for control information for link establishment and closure. S-frames (supervisory) carry numbered control information when the link is already established, e.g. requests for retransmissions or acknowledgments. The I+S-frames contain user information plus piggybacked supervisory information.

RLP uses a sliding window mechanism similar to TCP. The maximum window size is 61 which is also the recommended value. For multi-link operation (version 2) the maximum window size is extended to $496 - n \times (1 + T4/20\,\mathrm{ms})$, where n is the number of physical channels operating in parallel. $T4$ is the maximum allowed time difference of two frames when transmitted over different physical channels (see above, $T4 = 30\,\mathrm{ms}$ and $T4 = 50\,\mathrm{ms}$ for TCH/F9.6 and TCH/F14.4, respectively). For example, the recommended window size for 4 channels is 240.

RLP is a selective repeat ARQ protocol. If a transmitted frame suffers from errors, only this particular frame has to be retransmitted. This is achieved by the *Selective Reject* command. If an acknowledgment for a specific frame is not received by the transmitter for a given amount of time, the retransmission timer value, this frame is resent by the transmitter. The default retransmission timer value is 480 ms (TCH/9.6) or 520 ms (TCH/F14.4). The maximum number of retransmissions for a frame is set per default to 6. If the radio channel is in a very bad state, and frames have to be retransmitted a number of times, this could lead to remarkable delays. These delays could lead again to timeout problems on higher-level protocols (see Sect. 5.1 for further discussion on this problem).

RLP is always in one of two possible states: During connection establishment, connection termination and in idle state, RLP is in the *Asynchronous Disconnected Mode* (ADM). In this state the two endpoints are logically disconnected. Only unnumbered frames are allowed. When a connection is established, RLP stays in *Asynchronous Balanced Mode* (ABM). Now numbered information and supervisory frames can be sent.

3.1.3 Interworking Function (IWF)

In this section we will describe the interworking between the GSM network and the PSTN for data calls ([59]). Interworking for speech calls or to ISDN are not covered here. Interworking to the PSTN is classified into service, network and signalling interworking. For an already established data call, network interworking is required. It is involved every time an end-to-end bearer service between a mobile terminal and a terminal in PSTN or ISDN is required. Interworking scenarios for a number of possible bearer services are defined.

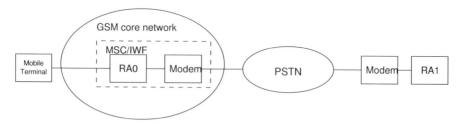

Fig. 3.6. Interworking for circuit-switched data between GSM terminal and PSTN [59]

The general interworking for circuit-switched data calls to the PSTN is outlined in Fig. 3.6. The Interworking Function (IWF) provides rate adaptation and a modem. The modem is connected to the PSTN and holds the connection to the counterpart modem on the fixed terminal at the PSTN. The IWF is located in an MSC. In the call setup the IWF selects the proper modem type with respect to the selected bearer capability. Another configuration, a *Shared Interworking Function* (SIWF), will be described below.

The rate adaptation process in the IWF is the reverse of that provided at the terminal adaptation in the mobile station (see Fig. 3.3 or Fig. 3.4). It converts the 64 kbit/s synchronous data stream at the A-interface between BSC and MSC into an asynchronous data stream suitable for the modem. The modem modulates the data stream for 3.1 kHz transmission in the PSTN. The GSM signalling is mapped to the ITU V-series functions. In the case of a non-transparent bearer service two data buffers exist. One buffer is required for transmission from MS towards fixed network for the case when the fixed network terminal is unable to accept the data. The second buffer is used for transmission from fixed network towards MS to prevent data loss when the radio path is temporary disconnected.

Shared interworking function (SIWF)
An IWF can be used only by its own MSC. A *Shared Interworking Function* (SIWF, [51]) can be used by several other network entities, e.g. a different MSC within the GSM core network. The reasons for establishing a SIWF are the following:

- To provide interworking functionality also in areas where the expected data traffic is low at no additional costs. The SIWF could be located in an area where the IWF functionality is needed anyway. The low traffic area MSC uses the SIWF just in case a (not very probable) data call is made. So not every MSC needs IWF functionality.
- If the number of data calls on a specific MSC/IWF exceeds the capabilities of the local IWF, an SIWF can handle the additional data calls. So the overloaded MSC can still serve the call although its IWF capabilities are already limited.

- New data services can be rolled out more quickly, because not every IWF needs to be upgraded instantaneously. At the beginning only a single SIWF needs to have the new service capabilities. All new services are routed over this SIWF.

The entity where the SIWF is located is called the SIWF server. The MSC using the SIWF functionality needs an additional entity, the SIWF controller (see Fig. 3.7). The SIWF server is connected via the K-interface to the MSC

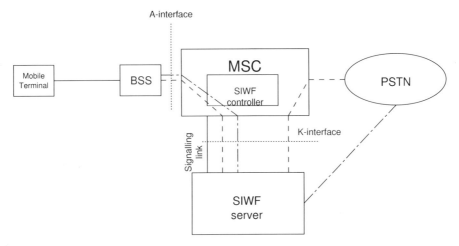

Fig. 3.7. Shared interworking function providing IWF functionality detached from a particular MSC [51]

responsible for the connection (*visited* MSC). The visited MSC routes the call to the SIWF server if in the visited MSC no IWF is available, or for other reasons (see above for SIWF motivation). For the routing from MS to PSTN there exist two possibilities: The dashed line indicates the first one, the *loop* method. The call is routed from the visited MSC to the SIWF server, back to the visited MSC (both times via the K-interface) and then routed to the PSTN by the visited MSC, as usual. The second possibility, the *non-loop* method is indicated by the dashed-dotted line in Fig. 3.7. The call is again routed from the visited MSC to the SIWF server but the SIWF provides immediate connection to the PSTN without any further MSC interaction. However, the second method is only applicable for mobile originating calls.

The SIWF provides the same functionality as a normal IWF plus additional routing and interface mechanisms necessary for signalling with the visited MSC and the PSTN. The physical location of the SIWF is not specified.

3.1.4 High-Speed Circuit-Switched Data (HSCSD)

High-Speed Circuit-Switched Data (HSCSD, [53, 54]) is a technique using up to 8 timeslots of the GSM radio interface simultaneously. In [54] explicitly up to 8 timeslots are allowed. Every timeslot provides a full-rate traffic channel, which leads to a maximum data rate for 8 TCH/F9.6 (TCH/F14.4) traffic channels of 76.8 (115.2) kbit/s. However, the A-interface restricts the user data rate to 64 kbit/s (see below and also [54]). As the term HSCSD already implies, the n channels are circuit switched, i.e. the resources are blocked for the entire call duration.

The logical architecture of HSCSD for a transparent bearer service is shown in Fig. 3.8. The number of parallel traffic channels (timeslots) is n

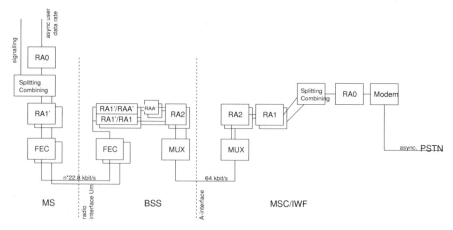

Fig. 3.8. Logical Architecture of High-Speed Circuit-Switched Data in the case of a transparent bearer service [129]

$(1 \leq n \leq 8)$. The user data stream is split at the MS and the MSC into n parallel channels. So all rate adaptation functions (RAx) except RA0 have to exist in multiple units. At the A interface (BSS to MSC) the n channels are multiplexed into a single 64 kbit/s. So the hard limit for HSCSD data rate is located at the A-interface in the core network and *not* at the radio interface! At the MSC this 64 kbit/s line is demultiplexed again for performing rate adaptation for n channels simultaneously. The rate adaptation functions are the same as in basic CSD operation.

Despite the splitting into n channels in different parts of the GSM network, the connection between MS and MSC is just a single logical one. This is also important for non-transparent bearer services. The RLP and L2R entities in the MS and the MSC operate *on top of* the splitting/combining functions. Thus there exist only one RLP (L2R) entity on either side of the reliable layer 2 connection (RLP). As the overall user data rate is much higher, the window size of the RLP protocol for HSCSD operation has been increased (see also Sect. 3.1.2).

HSCSD supports symmetric and asymmetric connections. *Symmetric* connections use n channels in both uplink and downlink. *Asymmetric* connections allocate the same resources as the symmetric ones. However, the mobile station uses only a subset of the allocated timeslots in the *uplink*. In the downlink direction all allocated timeslots are always used. This allows simpler mobile stations because the necessary sending capability is reduced. The limiting problem in the mobile station is the efficiency and thermal load on the power amplifier if more than one timeslot is used. The frames in the timeslots not used in the uplink direction are ignored by the IWF. Asymmetric radio interface connections are *only* applicable in non-transparent bearer services. In symmetric and asymmetric configurations one bi-directional channel always carries a Fast Associated Control Channel (FACCH) for signalling.

3.2 Packet-Switched Data, General Packet Radio Service (GPRS)

General Packet Radio Service (GPRS) is the packet-switched data service introduced in GSM Phase 2+ ([52]). Packet-switched services are much more suited to the bursty nature of data traffic than circuit-switched services. Resources are used only when there is data to transmit, otherwise they can be used for other services or users. GPRS allows packet switching in the complete mobile network including the radio interface. The new switching paradigm is reflected in an entirely new core network and protocol architecture, and also in a different radio interface protocol and mobility management. We will provide a brief introduction to the functionality of the parts essential in this book. Good overall introductions to GPRS are [36] and [30].

3.2.1 System Architecture

The simplified overall GPRS system architecture is shown in Fig. 3.9 [62]. Only the data path is shown, which we will concentrate on in the following. No subscriber management and operation management entities are included. The GPRS network is a new subsystem additional to the already existing subsystems in GSM.

To support packet switching within the mobile network two new elements are introduced. The *Serving GPRS Support Node* (SGSN) is the element serving the mobile station. It is connected to the base station subsystems via the *Gb*-interface. The SGSN is responsible for mobility management of GPRS enabled mobile stations and logical link control. It routes the packet to the right BSS when the MS is in its service area (those BSSs connected to this specific SGSN). Every SGSN serves a number of base station subsystems.

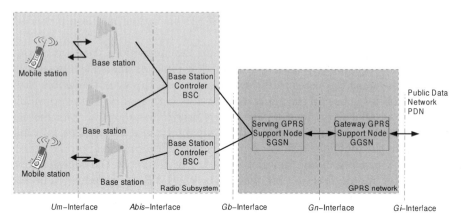

Fig. 3.9. Simplified system architecture of the General Packet Radio Service GPRS (data path only) [62]

The *Gateway GPRS Support Node* (GGSN) connects the GPRS core network to the external *Public Data Network* (PDN). The packets arriving from the PDN via the *Gi*-interface are tunneled to the proper SGSN. The GGSN evaluates the destination address in the packet and, together with maintained routing information, it sends the packet to the currently serving SGSN.

The upgraded GSM radio subsystem is reused in GPRS (see Sect. 3.2.3).

3.2.2 Services

GPRS bearer services are specified for end-to-end packet-switched data transfer. GPRS defines two service categories [61]:

- *Point to Point* (PTP) service: Packets are transferred between two dedicated users, initiated by the service requester and received by the receiver. The PTP service is offered in *connection-oriented* and *connectionless* mode. In connection-oriented mode a logical relation between the users exist that may last from seconds to hours. GPRS shall be able to support the ISO Connection Oriented Network Protocol X.25.
  In connectionless transfer mode *no* relation between consecutive packets exists. Every packet is sent completely independently of each other. In connectionless mode the Internet Protocol (IP) network layer protocol shall be supported.
- *Point to Multipoint* (PTM) service: PTM services support transfer of messages from a single user to multiple subscribers. Three different kinds of PTM services are defined:
  PTM-Multicast transmits messages to all subscribers within a geographical area(s) as defined by the service requester. No knowledge of receiver group members is necessary.

PTM-Group Call transmits messages to all subscribers within a group independently of their current geographical location. Instantaneous receiver group knowledge is required.

IP-Multicast is the multicast supported by the IP protocol suite itself. An IP-Multicast group can be distributed across the Internet. GPRS shall support IP-Multicast.

Service characteristics (Quality of Service)

Quality of Service (QoS) in packet-switched data transfer is defined in a different way to QoS in circuit-switched environments. In GPRS five QoS parameters are available ([63]), which are *service precedence, delay, reliability, peak throughput* and *mean throughput*. The user can select/negotiate the QoS parameters best suited for his actual application.

- The *Service Precedence* offers three classes of priority relative to other services.
- *Reliability* is defined in terms of residual error rates in the cases of data loss, out of sequence delivery, duplicate data delivery and corrupted data. The reliability classes are listed in Table 3.1. The negotiated reliability

Table 3.1. GPRS reliability classes in terms of residual errors for various error cases [63]

Reliability class	Probability for			
	packet loss	duplicate packet	out of sequence	corrupt packet
1	10^{-9}	10^{-9}	10^{-9}	10^{-9}
2	10^{-4}	10^{-5}	10^{-5}	10^{-6}
3	10^{-2}	10^{-5}	10^{-5}	10^{-2}

class subsequently specifies the used network layer protocol parameters. The combination of different modes of the GPRS network layer protocols (in various interfaces) supports the reliability requirements of Table 3.1. The combinations of the GPRS Tunneling Protocol (GTP), the Logical Link Control (LLC) layer and Radio Link Control (RLC) layer necessary for supporting a specific reliability class are listed in Table 3.2. For a description of the protocols see Sect. 3.2.4. GTP uses the TCP/IP protocol stack. For GTP acknowledged mode TCP and for GTP unacknowledged mode UDP is used. Signalling and SMS messages shall be transferred with reliability class 3.

- The *delay* class parameter specifies the maximum allowed mean delay and 95% delay allowed by the transfer of data through the GPRS network. The 95% delay is the maximum delay allowed in 95% of all data transfers. It is defined as end-to-end delay only *within* the GPRS network, i.e. from the mobile station to the *Gi*-interface at the GGSN. The specified delay classes are listed in Table 3.3. Delay class 4 defines a best effort service.

Table 3.2. GPRS reliability classes and the corresponding network layer protocol modes necessary to achieve the reliability class [63]

Reliability class	GTP mode	LLC frame mode	LLC data protection	RLC mode
1	Acknow.	Acknow.	Protected	Acknow.
2	Unacknow.	Acknow.	Protected	Acknow.
3	Unacknow.	Unacknow.	Protected	Acknow.
4	Unacknow.	Unacknow.	Protected	Unacknow.
5	Unacknow.	Unacknow.	Unprotected	Unacknow.

Table 3.3. GPRS delay classes. They specify maximum allowed delays within the GPRS network from mobile station to the Gi-interface [63]

Delay class	Maximum delay (s)			
	128 byte packet		1024 byte packet	
	Mean delay	95% delay	Mean delay	95% delay
1 (predictive)	< 0.5	< 1.5	< 2	< 7
2 (predictive)	< 5	< 25	< 15	< 75
3 (predictive)	< 50	< 250	< 75	< 375
4 (best effort)	unspecified			

Packets are sent when the resources are available, however, the user has no guarantee of the maximum delay the packet is allowed to have.

• The *throughput* specifies the expected bandwidth the user requires for a transfer. It is defined in *mean* and *peak* throughput classes. There is no guarantee that the peak throughput negotiated at the beginning is ever reached. Only the mean throughput must be maintained during the lifetime of a connection. The mean throughput may be limited by the network even if more resources were available. It is measured in bytes per hours while the peak throughput is measured in bytes per second. Table 3.4 lists the defined peak and Table 3.5 the mean throughput classes. Note that the peak throughput classes are defined up to 2 Mbit/s, while the GPRS radio interface only supports a maximum of 171.2 kbit/s user data rate. The higher classes are already defined for EDGE[2] and the third-generation system UMTS (see Chapter 4).

3.2.3 Radio Interface

GPRS uses the same resources as the basic GSM system, i.e. the FDMA-TDMA radio interface. A resource unit is a timeslot-frequency pair. However, GPRS allows multislot operation within a single TDMA frame served from a

---

[2] *Enhanced Data rates for GSM Evolution* (EDGE) introduces a new modulation scheme at the GSM radio interface. GMSK is replaced by 8PSK if good radio channel quality is available.

Table 3.4. GPRS peak throughput classes [63]. This is the maximum throughput that can be achieved in a transfer when a specific class has been negotiated

Peak throughput class	Peak throughput (byte/s)
1	up to 1 000 (8 kbit/s)
2	up to 2 000 (16 kbit/s)
3	up to 4 000 (32 kbit/s)
4	up to 8 000 (64 kbit/s)
5	up to 16 000 (128 kbit/s)
6	up to 32 000 (256 kbit/s)
7	up to 64 000 (512 kbit/s)
8	up to 128 000 (1024 kbit/s)
9	up to 256 000 (2048 kbit/s)

Table 3.5. GPRS mean throughput classes [63]. This is the mean throughput achieved during the entire lifetime of a connection

Mean throughput class	Mean throughput (byte/h)
1	up to 100 ($\approx$0.22 bit/s)
2	up to 200 ($\approx$0.44 bit/s)
3	up to 500 ($\approx$1.11 bit/s)
4	up to 1 000 ($\approx$2.2 bit/s)
5	up to 2 000 ($\approx$4.4 bit/s)
6	up to 5 000 ($\approx$11.1 bit/s)
7	up to 10 000 ($\approx$22 bit/s)
8	up to 20 000 ($\approx$44 bit/s)
9	up to 50 000 ($\approx$111 bit/s)
10	up to 100 000 ($\approx$0.22 kbit/s)
11	up to 200 000 ($\approx$0.44 kbit/s)
12	up to 500 000 ($\approx$1.11 kbit/s)
13	up to 1 000 000 ($\approx$2.2 kbit/s)
14	up to 2 000 000 ($\approx$4.4 kbit/s)
15	up to 5 000 000 ($\approx$11.1 kbit/s)
16	up to 10 000 000 ($\approx$22 kbit/s)
17	up to 20 000 000 ($\approx$44 kbit/s)
18	up to 50 000 000 ($\approx$111 kbit/s)
31	Best effort

mobile station. Also resource units can be allocated for uplink and downlink separately. The packet-switching principle is taken into account also at the radio interface: GPRS channels allocate resources only for the time that data is ready to transfer and releases the resources immediately when they are not needed. This results in efficient resource usage, multiple users can use the same resources in time multiplex.

GPRS introduces four **channel coding** schemes which allow for different user data rates per timeslot. The coding schemes CS-1 to CS-3 use the same convolutional code but different puncturing levels, which results in different code rates and thus protection quality. CS-1 additionally has a 40-bit block fire code. CS-4 has no coding at all. Thus CS-1 has the lowest user data rate but the best error protection. Coding in GPRS is always done for a single RLC/MAC radio block which always has a coded length of 456 bits. This block is interleaved over 4 normal bursts, each covering 114 data bits. Hence the interleaving is much more reduced compared to circuit-switched data (interleaving over 19 bursts). The radio block length before coding depends on the coding scheme and varies from 181 to 428 bits. The coding schemes and achievable net user data rates for a few multislot combinations are summarized in Table 3.6. For coding of signalling channels CS-1 is always used.

Table 3.6. GPRS coding schemes and net throughput for different number of allocated timeslots (TS)

Coding scheme	Code rate	Uncoded radio block, payload (bit)	Coded bits (bit)	Punc-tured bits (bit)	Data rate 1 TS	2 TS	4 TS	8 TS
						(kbit/s)		
CS-1	1/2	181	456	0	9.05	18.1	36.2	72.4
CS-2	$\approx 2/3$	268	588	132	13.4	26.8	53.6	107.2
CS-3	$\approx 3/4$	312	676	220	15.6	31.2	62.4	124.8
CS-4	1	428	456	0	21.4	42.8	85.6	171.2

The GPRS **radio interface protocol** is related to the physical layer and the RLC/MAC layer of the GPRS protocol architecture. The principles include a master-slave concept and the capacity-on-demand principle.

The *master-slave* concept is related to the logical channel assignment onto the physical channels. The control channels are mapped only to a single physical packet data channel (PDCH) located on a single timeslot acting as master. The other PDCHs (on the other timeslots) are only used for user data transfer. These are acting as slaves.

The allocation of PDCHs is based on the *capacity-on-demand* principle. As the radio resources are shared with circuit-switched data and speech users, a resource management unit regulates the access to the radio resources. Thus usually not all resources are available to GPRS traffic. The operator has to decide which traffic has precedence. The number of PDCHs in a cell can be changed according to available resources and the offered traffic load. If, for example, a mobile station wants to transfer more data than is possible on the currently allocated resources, it sends a message on a control channel to the network. The network decides whether the resources are granted or not, depending on the current network load and other parameters like negotiated service class. A load supervision function, typically located in the MAC layer,

may monitor all radio resource specific parameters and builds up the basis
for the decisions.

In GPRS multiplexing on the radio interface is based on RLC/MAC packets (see also Sect. 3.2.4). Every message is included in a RLC/MAC packet,
which occupies at least 4 timeslots, a *radio block*. The timeslot occupancy
is defined in the 52-multiframe, framing 52 timeslots to a multiframe. These
52 timeslots are *not* consecutive in time but rather every n^{th} timeslot in a
GSM frame. See Fig. 3.10 for illustration. Thus there exist 8 multiframes, for
each of the 8 timeslots in a GSM frame. A radio block is transmitted on 4
timeslots of a single multiframe. Thus the transmission of a single radio block
lasts exactly 4 GSM frames ($4.616\,\mathrm{ms}\times4 = 18.464\,\mathrm{ms}$).

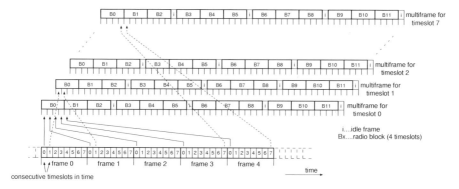

Fig. 3.10. Multiplexing of the RLC packets and timeslots in the 52-multiframe in
GPRS

The capability of mobile stations to send on multiple timeslots per TDMA
frame is defined as *multislot capability*, which are classified in *multislot classes*.
The multislot classes and the number of multiple timeslots for transmission
and reception are defined in [68]. Currently 29 different multislot classes are
defined. There exist, furthermore, two different *types*, Type 1 and Type 2.
Mobile stations belonging to multislot classes of Type 1 are not required to
transmit and receive at the same time. Type 2 mobile stations are required
to transmit and receive simultaneously.

3.2.4 Protocol Architecture

The protocols can be logically divided into protocols for data transmission
and protocols for signalling. We will focus only on the transmission protocols
in this book.

Figure 3.11 shows the GPRS protocol architecture of the transmission
plane [63]. Protocol stacks for mobile station (MS), base station subsystem

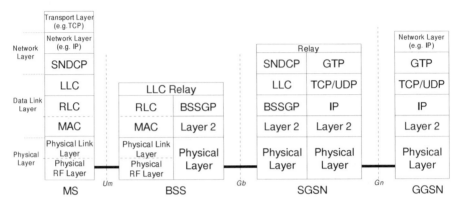

Fig. 3.11. GSM GPRS protocol stack of the transmission plane (data path)

(BSS), serving GPRS support node (SGSN) and gateway GPRS support node (GGSN) are shown. The counterpart of the transport layer at the MS is at the endpoint of the complete connection, which is not shown here. Below the transport layer, the network layer provides the basic datagram service functions. GPRS supports currently IP and X.25 as network layer protocols. Both transport layer and network layer are beyond the scope of the GPRS specification.

GPRS Tunneling Protocol (GTP) . In the GPRS backbone (*Gn*-interface between SGSN and GGSN) the GPRS Tunneling Protocol (GTP) is used to tunnel the network layer packets between the two GSNs. It is specified in [60]. During tunneling, the original packets are encapsulated within a new packet and thus they are not modified. Below GTP, a TCP/IP or UDP/IP backbone is used as transmission protocol. An Ethernet, ISDN, or ATM transmission method can be used below the tunneled IP layer.

Subnetwork-Dependent Convergence Protocol (SNDCP) . The upper layer protocol between the MS and the SGSN is the Subnetwork-Dependent Convergence Protocol (SNDCP) [67]. SNDCP provides transmission of the network layer protocol data units (N-PDUs) in a transparent way. There are 4 main functions:

- Multiplexing of several Packet Data Protocols (PDPs). PDP denotes the upper layer protocol (i.e. IP or X.25). SNDCP is able to transmit simultaneously packets from different network layer protocols. They are distinguished by an identification field in the protocol header.
- Compression/Decompression of user data. Data compression is an optional feature. The compression algorithm is negotiated between the two protocol endpoints. Currently ITU-T V.42bis compression is supported.
- Compression/Decompression of protocol control information. The network layer protocol header is optionally compressed. The compression algorithm

used is negotiated between the protocol endpoints. Van Jacobson TCP/IP header compression is available.

- Segmentation of network layer packets (N-PDUs) into Logical Link control (LLC) packets and reassembling of LLC packets into N-PDUs. The segmentation is done according to the maximum LLC packet length (see below).
- Buffering of N-PDUs at the SGSN.

It is assumed that the underlying LLC protocol has acknowledged and unacknowledged data transfer mode. Accordingly, SNDCP can be configured for acknowledged or unacknowledged transmission. For transmission SNDCP selects the corresponding LLC mode. So despite SNDCP having an acknowledged mode, SNDCP itself does *not* provide reliability but uses the functions of LLC. The header of the SNDCP layer consists of 2, 4 or 5 bytes. This depends upon whether segmentation is switched on (actual segment offset has to be stored) and the range of the N-PDU number. The length of the packets is variable, but the LLC layer sets an upper limit.

Logical Link Control (LLC). The Logical Link Control (LLC) layer [66] is part of the data link layer. It spans from the MS to the SGSN and provides packet transmission between these two entities. LLC provides a highly reliable logical link independently of the underlying radio interface protocol. There are two transfer modes (see Fig. 3.12):

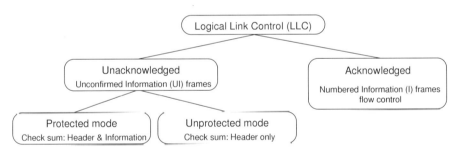

Fig. 3.12. Illustration of Logical Link Control modes

- In *unacknowledged mode* Unconfirmed Information (UI) frames are transmitted. There is no error recovery nor reordering of packets included. However, a check sum detects transmission errors and also frame format errors are caught. Packets in error are discarded.
  The check sum could protect header and information field (protected mode) or only the header field (unprotected mode).
- In *acknowledged mode* SNDCP packets are transmitted in numbered Information (I) frames. Error recovery and retransmissions of packets in error is provided within LLC. Acknowledged mode requires the establishment of the Asynchronous Balanced Mode (ABM) at the beginning.

For transmission LLC sends frames with sequence numbers and the receiver acknowledges with an ACK or selective ACK. If a packet gets lost, either the transmitter gets a retransmission timeout or it receives an acknowledgment for one of the following packets. Then the transmitter knows the packet is either missing or delivered out of order. The transmitter then could explicitly request an ACK for this packet or retransmit it immediately. The maximum number of retransmissions is set to 3. The maximum number of outstanding frames is between 2 and 16, depending on the service access point used (i.e. required quality of service).

In ABM mode flow control functions are defined. The receiver can stop the data flow temporarily when it is busy by sending a message to the other peer.

If the error is unrecoverable a notification is sent to the upper layer and the management unit.

When establishing a connection a temporary logical link identifier (TLLI) is issued. LLC supports multiple TLLI sessions simultaneously, all distinguished by their TLLI. Also the GPRS mobility management functions and the SMS entities use the LLC layer for transfer of information each multiplexed by LLC.

The LLC frame format is drawn in Fig. 3.13. The address field includes the

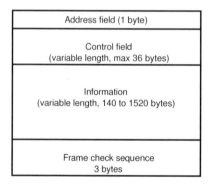

Fig. 3.13. LLC frame format including header, control, information and frame check sequence field

service access point identifier to distinguish the Service Access Point (SAP) used and a Command/Response field for distinction between command and response frame. It is one byte long.

The control field contains a special entry for each type of frame (I for confirmed information transfer, S for supervisory functions, UI for unconfirmed information transfer and U for control functions). All necessary control information is stored in this field, as there are sequence numbers, state variables, a bitmap for selective acknowledgment identifying the missing segments, encryption mode, and the like. The length depends on the frame format and is at maximum 36 bytes.

Packet payload is stored in the variable length data field. It could last from 140 to 1520 bytes. The default value is 1503 bytes.

The last field contains the Frame Check Sequence (FCS). It is a 24-bit cyclic redundancy check (CRC) code. It detects errors in both header and data fields of the LLC frame. The exception is the unprotected mode of the unacknowledged mode, where the CRC is computed only over the header entries.

The LLC layer also provides user data confidentiality by means of *ciphering*. Also user identity confidentiality is supported. A 64-bit ciphering key is used as input for the ciphering algorithm.

Radio Link Control/Medium Access Control (RLC/MAC). The lower part of the data link layer is divided into the Radio Link Control (RLC) and into the Medium Access Control (MAC). Both protocols belong to the *Radio Resource Management* and are defined in [65].

The RLC layer protocol provides the following functions:

- Segmentation of LLC packets into RLC data blocks and reassembling at the corresponding endpoint.
- Segmentation of RLC/MAC control blocks into RLC/MAC blocks.
- Providing a backward error correction for establishing a reliable link.

The RLC block size is constant for a specific channel coding scheme (see Table 3.7). The structure of a RLC/MAC block is shown in Fig. 3.14. A

Table 3.7. RLC data block size for the four GPRS coding schemes. No MAC header is included here

Coding scheme	RLC data block (bytes)
CS-1	22
CS-2	32 7/8
CS-3	38 3/8
CS-4	52 7/8

RLC/MAC block consists of a MAC header which has fixed size of 1 byte, a variable length RLC header, an information field and spare bits. For transmission of RLC/MAC control messages the RLC header is replaced by a *control header* and the RLC information field together with the spare bits are replaced by the RLC/MAC signalling information field.

The RLC protocol can operate in both acknowledged or unacknowledged mode. In acknowledged mode, RLC provides a reliable link (using the ARQ technique) between the mobile station and the base station. This is different from GSM circuit-switched Radio Link Protocol (RLP). RLP spans from

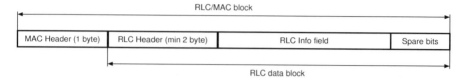

Fig. 3.14. Structure of a RLC/MAC block including a MAC header (fixed length), a RLC header (variable length), an information field and some spare bits

MS to the MSC, whereas RLC/MAC in GPRS spans only from MS to the BSS. RLC uses acknowledgment packets to acknowledge the data packets. Selective retransmission is supported, thus only the packets in error have to be retransmitted. A sending window is maintained (similar to TCP and RLP) which is 64 blocks wide. Thus a maximum of 64 packets can be outstanding still unacknowledged. The in-order-delivery is maintained by block sequence numbers which have a range from 0 to 127.

In unacknowledged mode ACK packets are also sent, however, no retransmissions are ever performed. The ACK/NACK packets are only for control and signalling. Thus one has to be aware that backward channel capacity is also occupied by ACK packets although the unacknowledged mode is active.

The **Medium Access Control** (MAC) layer controls the management of the shared transmission resources. It supports point-to-point *Temporary Block Flows* (TBF) which is a physical connection between two RLC/MAC endpoints. A TBF is allocated a radio resource on one or more Packet Data Channels (i.e. timeslots). It is maintained only for the duration of the data transfer. For establishment of a TBF a random access procedure is performed. The random access uses a slotted ALOHA protocol.

The resource allocation can by dynamic or fixed. The resource allocation is always granted from the network. For dynamic allocation of uplink resources *Uplink State Flags* are used. They are transmitted from the base station to the MS and indicate the resources the MS is allowed to use in the uplink.

Scheduling of incoming traffic to the allocated resources is also done by the MAC layer. RLC/MAC acknowledgment packets always have higher priority than data packets. However, the detailed scheduling is up to the implementation and not specified within the recommendation.

Protocol overhead. The total protocol stack from TCP down to RLC MAC is illustrated in Fig. 3.15 The data packet from the layer on top of TCP (typically the application layer) is framed and segmented into smaller blocks. Every protocol adds its own header, which increases the protocol overhead significantly. A quantitative evaluation is plotted in Fig. 3.16. Figure 3.16 shows the overhead *per* data packet, absolute overhead in Fig. 3.16(a) and relative values in Fig. 3.16(b). We assumed *no* IP fragmentation in here. So the TCP/IP header overhead may be higher if packets travel over a path, where they get segmented in the IP layer. The total GPRS overhead is com-

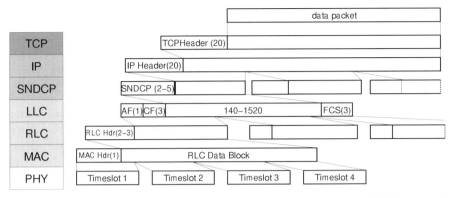

Fig. 3.15. Illustration of the packet-framing procedures done in the GPRS protocol stack

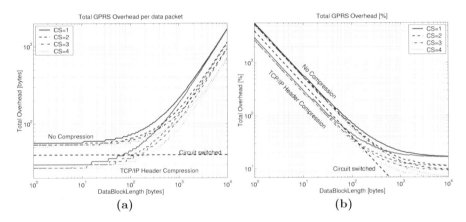

Fig. 3.16. Evaluation of the total GPRS protocol overhead per data packet for varying sizes of the input data packets (from protocol on top of TCP). Absolute values are plotted on the left-hand side plot **(a)**, relative values in the right-hand side **(b)**. See also Fig. 3.15

puted for various data packet sizes. A data packet is the topmost packet delivered to the TCP protocol from a higher-layer protocol. In general, total overhead grows for increasing data packet size. Because a larger data packet is split into more smaller LLC and RLC/MAC blocks, more headers are needed which increases the overhead. We plotted the overhead for all four GPRS channel coding schemes with and without TCP/IP header compression. Without TCP/IP header compression, the total overhead starts between 50 to 60 bytes for a small data packet size. If the data packet size exceeds the limits for segmentation of LLC and RLC layers the overhead starts to increase significantly. This leads to a constant relative overhead for high data block sizes (Fig. 3.16(b)).

With TCP/IP header compression, overhead starts at approximately 30 bytes per data packet which is only about 50% of the value when omitting header compression. However, for larger data packet sizes, the difference between the two cases, with and without header compression, diminishes.

For comparison we also included the overhead for circuit-switched data. It does *not* depend on the data packet size. It remains constant at about 40 bytes, representing the 20-bytes TCP and 20-bytes IP header. However, the relative overhead is, of course, vanishing as data packet size increases. Note that we assumed no IP segmentation here. If we include IP segmentation, circuit-switched data overhead would increase as well with data packet size. But this is also true for GPRS, so the relative difference between GPRS and CSD will remain.

For large data packet size the different channel coding schemes lead to significantly different overhead (right-hand side in Fig. 3.16). The differences almost diminishes for small data packet size.

From Fig. 3.16 we observe that increasing data block size does *not* decrease the overhead necessarily. In GPRS relative overhead per data packet remains constant independent of the data packet size (above a specific threshold data packet size). This is different from CSD, where the relative overhead decreases asymptotically with growing data packet size. This makes it much more difficult to choose an optimum data packet size for transmission.

3.3 Link Level Simulations

Link level in this context is a connection on the *physical layer* between the mobile station and base station. This is *not* a connection in the data link layer of the OSI model. In this section we derive the performance of the physical layer of the GSM radio interface by means of simulation. The performance measure is Bit Error Ratio (BER). We will provide results for GSM-CSD as well as GPRS traffic channels. These will be necessary for further investigations of the upper protocol layers.

The physical layer procedures of the GSM link level are outlined in Fig. 3.17. The GSM link consists of forward error correction (FEC), interleaving, burst formatting, GMSK modulation, a channel model for modeling of the radio interface, GMSK receiver, deinterleaving and FEC decoding. We implemented the whole physical layer chain in Ptolemy [127], a simulation tool for heterogenous simulation models. Together with the COST 207 channel models, realistic results for link level performance could be obtained.

The **forward error correction** and **interleaving** part depends on the traffic channel (see left-hand side of Fig. 3.17). Because the air interface rate is constant for all traffic channels, the channel coding quality depends on the user data rate. The other parts, burst formatting, modulator, demodulator and the channel model, are the same for each traffic channel.

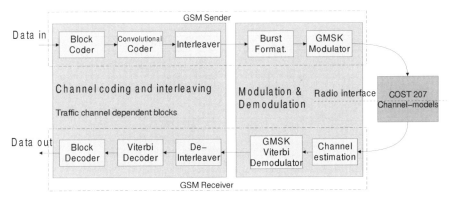

Fig. 3.17. Physical layer functions of a GSM transmitter and receiver. The radio interface is modeled by the COST 207 channel models

3.3.1 Implementation Details

In this section we will present the detailed implementation of the link level simulator. We will start at the data input of the GSM sender (upper left part of Fig. 3.17).

The FEC blocks in the transmitter (block and convolutional encoder and interleaver) are built straightforwardly according to the specification in [58], there are no performance relevant details. The block coder for GPRS PDTCH CS-1 coding scheme is defined by a generator polynomial ([58]). For all other coding schemes the block coder degenerates and only adds 4 tail bits.

The **GMSK modulator** sends GMSK pulses for every bit. A GMSK pulse with 3 dB bandwidth-bit period product of BT = 0.3 is plotted in Fig. 3.18. A bit duration in GSM is $T = 3.692\,\mu s$. The upper curve is the impulse response, $g(t)$, of the Gaussian filter used in GMSK modulation. GMSK modulation has a constant envelope but a variable phase. A GMSK pulse is, by definition, infinite in time. In the simulation it is cut to a total length of 6 bit durations (as shown in Fig. 3.18). Then the settling time at both ends of the pulse is long enough to eliminate distortions. The pulses are generated with time resolution of $T_c = T/\eta$, where η is the oversampling factor. When sending a bit sequence, it is NRZ (Non Return to Zero) encoded, a positive (+1) or negative (−1) GMSK pulse is sent for each bit. As a GMSK pulse has a duration of more than a single bit period, overlapping between the GMSK pulses occurs. The sequence of overlapped GMSK pulses is integrated and corresponds directly to the phase of the GMSK modulated bit sequence. A sample bit sequence is shown in Fig. 3.19.

The GMSK modulated sequence is fed into the radio channel. The radio channel is an equivalent baseband model which is described in detail in Sect. 3.3.2.

Now we turn to the receiver part of the GSM link (lower half in Fig. 3.17, starting from right to left). We start at the demodulator. Baseband process-

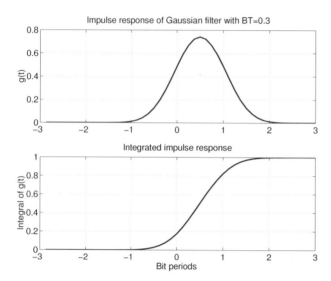

Fig. 3.18. GMSK pulse with bandwidth-bit period product of BT $= 0.3$. The lower curve corresponds to the integrated GMSK pulse

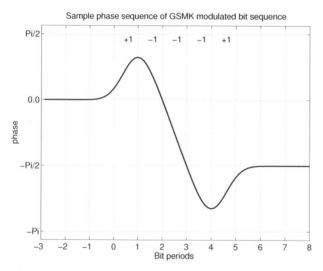

Fig. 3.19. Sample phase sequence of GMSK modulated bit sequence. The bit sequence is shown in the upper part of the figure

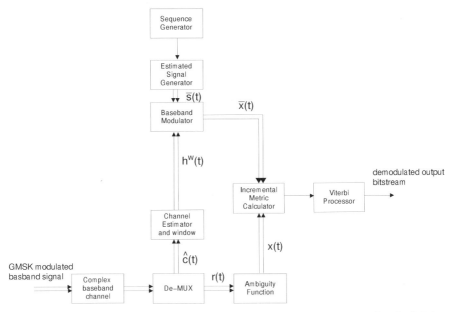

Fig. 3.20. Structure of the baseband processing in the demodulator [123]. ©John Wiley & Sons Limited. Reproduced with permission

ing of the demodulator is illustrated in Fig. 3.20. The theoretical background can be found in [123]. Only a short description is given in the following: The signal coming from the complex baseband channel is demultiplexed. Demultiplexing here means extraction of the training sequence out of the burst. The extracted training sequence, $\hat{c}$, is used to estimate the channel impulse response, h (h is not shown explicitly in Fig. 3.20). The **channel estimator** uses the properties of the known training sequence (the central 26 bits of a GSM burst). It correlates the training sequence bits of the received burst with the ideal one. Thus an estimate of the channel impulse response is available. Also the timing offset of the received burst is obtained. The estimated channel impulse response, h, is windowed with a rectangular window which truncates the impulse response. The final impulse response is denoted therefore by h^w.

A sequence and signal generator creates all possible GMSK phase sequences, $\bar{s}$, with length l (measured in symbol durations, T), the equalizer length. The sequences, $\bar{s}$, are generated off-line at the beginning and stored in memory. For every new channel estimate (i.e. every GSM burst), the estimated channel impulse response, h^w, is then used to create distorted versions of all possible signal combinations, $\bar{x}$. They are computed by convolution of $\bar{s}$ with the estimated impulse response, h^w. Afterwards, the signal without training sequence, r, is multiplied by the ambiguity function. Subsequently, the sequence $x(t)$ is compared with all possible signal combinations $\bar{x}$ to create the metric values for all these combinations. Finally the **Viterbi processor** finds the maximum likelihood sequence (MLSE).

For decoding of the convolutional code another **Viterbi processor** is implemented. Our version provides hard decisions at the output. We took this kind of decoder because it deals in a natural way with punctured bits. Also, the achieved performance is state of the art.

The **block decoder** is only relevant in the case of the PDTCH CS-1 coding scheme in GPRS. In all other data traffic channels and coding schemes the block code is degenerated to adding/removing 4 tail bits. For PDTCH CS-1 a shortened binary cyclic code (Fire code) is used. It is especially designed to combat burst errors. The code is systematic and includes 40 parity bits. It can correct single burst errors of length $l = 12$. For decoding and error correction we used an 'error-trapping decoder' ([91, 116]).

The performance of the Fire block code, the convolutional code and the concatenated coding system of the GPRS PDTCH CS-1 traffic channel is shown in Fig. 3.21. This evaluation is without any modulation/demodulation and burst formatting, only the performance of the code stand-alone on an additive white Gaussian noise (AWGN) channel. The code performance is mea-

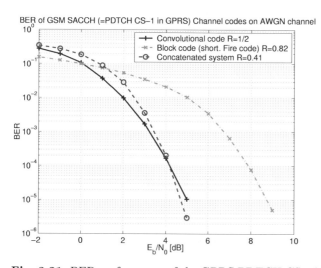

Fig. 3.21. BER performance of the GPRS PDTCH CS-1 block code, convolutional code and the concatenated coding system on an AWGN channel. This is the plain code performance without any modulation/demodulation and burst formatting. R is the code rate

sured in Bit Error Ratio (BER) over the signal-to-noise ratio SNR (Eb/N_0). Figure 3.21 demonstrates that the convolutional code contributes the main part to the performance of the concatenated coding system. Only for very low BER $< 10^{-5}$ is there an advantage in using both codes.

3.3.2 Channel Model

As radio channel model we used the COST 207 wideband channel model [96]. The channel model is a tapped delay line model and implemented as equivalent baseband model. There exist 4 different area types (propagation classes), 'Typical Urban' (TU), 'Bad Urban' (BU), 'Rural Area' (RA) and 'Hilly Terrain' (HT). Each class is characterized by its typical power delay profile (PDP). These PDPs were derived from a number of measurement campaigns in several European cities. For each area type a number of power-delay taps are defined which represent the multi-path propagation of radio signals in mobile radio channel environment. The taps model the power-delay profile of a received signal. Fading is added by a corresponding Doppler spectrum for each delay tap. The delay taps for all 4 area types are shown in Fig. 3.22. If the delay of particular taps is longer than the symbol period of

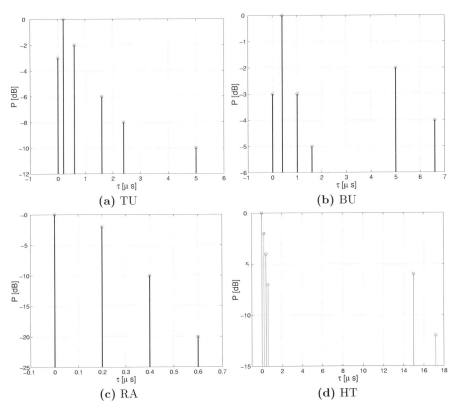

Fig. 3.22. Delay taps of the COST 207 channel model for 'Typical Urban' (**a**), 'Bad Urban' (**b**), 'Rural Area' (**c**) and 'Hilly Terrain' (**d**) area type. For each delay tap a corresponding Doppler spectrum is also defined for contribution of fading [96]

$T = 3.692\,\mu$s inter-symbol interference (ISI) is introduced. Thus an equalizer in the receiver is required.

The complex output samples, s_{out}, at the sampled time, t, is

$$s_{\text{out}}(t) = \sum_{k \in D} s_{\text{in}}(t - k) p_k r_k \exp(\text{j}2\pi f \tau_k)\,, \tag{3.1}$$

where $s_{\text{in}}(t - k)$ is the input sample at time $t - k$, r_k is the filtered Rayleigh coefficient for sampled delay k, f is the carrier frequency, p_k is the mean power for sampled delay time k and τ_k is the delay in seconds for the sampled delay time k. The summation is done over all taps.

The filtered Rayleigh coefficients r_k contribute to the instantaneous power of each delay tap and model small-scale fading. They are computed for each delay tap independently, as described in the following: A Rayleigh distributed random number sequence is filtered in a FIR filter. The frequency response of the filter is defined by the Doppler spectrum of the delay tap ([96]). Four Doppler spectra are defined in [96], classical Jakes spectrum, two different Gaussian spectra, and a Rice spectrum. The relation between taps and Doppler spectra is fixed according to the channel model parameters. The Doppler spectra vary according to the maximum mobile station velocity, v. This corresponds to the maximum Doppler frequency shift, $f_{\text{D,max}} = f \times v/c$. c denotes the velocity of light.

Because implementation is discrete in time, the delay taps of the model (Fig. 3.22) must be matched to the grid of sampled time $T_c = T/\eta$ with oversampling factor η. Here, it is important that η is chosen reasonably high to avoid significant matching errors. For example assume $\eta = 3 \rightarrow T_c = 1.23\mu$s. The maximum resolution in the simulation is thus T_c. Each delay tap has to be shifted (or merged) to the next point in the T_c grid. Here the second delay tap after the main tap could occur at the earliest time T_c after the main tap. This would corrupt the result significantly. Therefore we always used an oversampling factor of $\eta = 8$ in our simulations, which is a good compromise of reducing simulation run time and minimization of the error induced.

Summarizing the parameters of the channel models:

- Carrier frequency, f.
- Area type, one out from TU, BU, RA, HT.
- Mobile station velocity, v.
- Oversampling rate, η, (in this book always fixed to $\eta = 8$).

For modeling of a co-channel interferer, a second (simplified) GSM transmitter is used. It consists of a random sequence generator, burst formatter and a modulator. The modulated baseband bit stream is sent over the channel model and added to the user bit stream with the specified carrier-to-interference ratio, C/I, and signal-to-noise ratio, S/N. S/N can be set for user and interferer signal separately. The GSM transmitter for actual data transmission and the interferer are synchronized in time. Only a single co-channel interferer is modeled. No adjacent-channel interferer is included.

3.3.3 Simulation Results

In this book we have made simulations mainly using the 'Typical Urban' model. For 'Bad Urban' and 'Hilly Terrain' only results of the uncoded GSM link (without channel coding) are given. First we will show the performance of the raw, uncoded link and, then, of the link with full forward error correction included.

Uncoded GSM link. The uncoded link includes burst formatting, modulation/demodulation and channel model (only the right-hand side of Fig. 3.17). Channel coding and interleaving is omitted here. The evaluation of the raw uncoded GSM link has been done separately for S/N and C/I. For the S/N simulations (varied S/N) the C/I was fixed to $C/I = 100\,\mathrm{dB}$. Also, for the C/I simulations (varied C/I) the S/N_{user} and the S/N_{intf} were fixed to $S/N_{\mathrm{user}} = S/N_{\mathrm{intf}} = 100\,\mathrm{dB}$. The simulation parameters for the uncoded GSM link are listed in Table 3.8. The number of simulated bits varies accord-

Table 3.8. Simulation parameters for the uncoded GSM link

Parameter	Value (or range)
Velocity (km/h)	10,50,100,150,250
Carrier frequency (MHz)	903
Channel type	COST 207 or AWGN
Channel area type	'TU','BU','HT'
Oversampling rate	8

ing to the achieved BER. At least 2000 bit errors must have occurred to stop the simulation.

BER performance of the uncoded link for varied user S/N and varied C/I is plotted in Fig. 3.23. Velocity of the mobile station was set to 10 km/h, 50 km/h, 150 km/h and 250 km/h. The BER for COST 207 TU channel for $S/N > 8\,\mathrm{dB}$ and $C/I > 25\,\mathrm{dB}$ and with $v = 10$ km/h is almost two orders of magnitude below the BER for $v = 250$ km/h. BER for other mobile station velocities lie in between. A velocity of $v = 50$ km/h hardly degrades performance compared to $v = 10$ km/h. However, a big step can be observed when going up to $v = 150$ km/h. Then BER grows about one order of magnitude.

An error floor exists even at high S/N and C/I values due to the mobile radio channel. The error floor for high C/I values varies between 3×10^{-4} for $v = 10$ km/h and 8×10^{-3} for $v = 250$ km/h. For comparison the AWGN channel performance is plotted. BER on the AWGN channel does not have an error floor for high C/I and S/N values.

Two main effects are responsible for this irreducible error floor: Frequency selective fading caused by multi-path time delay spread causes inter-symbol interference (ISI). If the equalizer (in this simulation a MLSE Viterbi equalizer) is not able to combat the ISI entirely, an error floor results. Secondly,

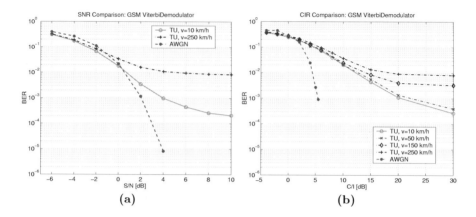

Fig. 3.23. BER of uncoded GSM link over AWGN and COST 207 typical urban ('TU') channel as a function of user S/N **(a)** and C/I **(b)**. The mobile station movement was set to 10 km/h, 50 km/h, 150 km/h and 250 km/h

Doppler spread due to mobile station or scatterer velocity introduces a random frequency modulation. This again results in an irreducible error floor. Details can be found in [113, 123].

The raw, uncoded BER performance of the demodulator for bad urban ('BU') and hilly terrain ('HT') channel model are plotted in Fig. 3.24. BER for BU channel model is almost one order of magnitude below that for TU (compare Fig. 3.23(b)). This is caused by the difference of the channel impulse responses (compare Fig. 3.22(a) and Fig. 3.22(b)). The BU channel model

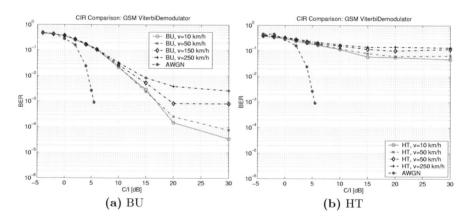

Fig. 3.24. BER of uncoded GSM link over AWGN and COST 207 bad urban ('BU') channel **(a)** and hilly terrain ('HT') channel **(b)** as a function of C/I. The mobile station movement was set to 10 km/h, 50 km/h, 150 km/h and 250 km/h. Note the different scaling on the ordinate

consists of a relatively strong delay tap at 5 µs. This has a certain 'diversity' effect, i.e. the equalizer can capitalize on this energy and the second dominant path is used, too. In the TU model only a single dominant path (the area from 0 s to 1 µs) is available. The probability that both dominant paths are in a fading dip is much lower than if only a single dominant path exists. Thus performance with the BU model is better than when using the TU model.

But if the excess delay between the two dominant paths exceeds a specific value, the second path can *not* be equalized any longer and introduces severe ISI. The maximum delay is given by the equalizer. The equalizer is able to equalize an ISI-corrupted signal only up to a maximum delay, which is given by the equalizer length. This effect can be observed for the hilly terrain ('HT') channel model. The second delay tap group has an excess delay of about $\tau = 15$ µs. The equalizer is *not* able to combat this severe ISI any longer, thus BER performance is very poor (see Fig. 3.24(b)).

Full GSM link. Now we present the performance of the fully encoded link. This includes channel coding and interleaving (entire Fig. 3.17). The results depend, of course, on the used channel coding scheme, i.e. on the traffic channel.

In Fig. 3.25, we plot BER over S/N and C/I for the circuit-switched data traffic channels TCH/F9.6 and TCH/F14.4 for various velocities. Due

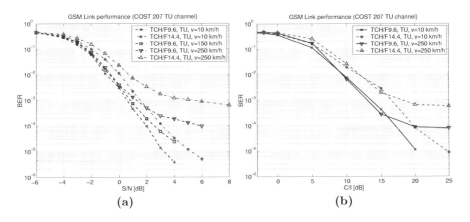

Fig. 3.25. BER of coded GSM link over COST 207 typical urban ('TU') channel as a function of user S/N **(a)** and C/I **(b)**. The mobile station movement was set to 10 km/h and 250 km/h. TCH/F9.6 and TCH/F14.4 traffic channel coding are shown

to the worse channel coding for TCH/F14.4, the BER is almost exactly one order of magnitude worse than for TCH/F9.6. This is true for both S/N and C/I variations. Because current cellular mobile radio networks are interference limited, the C/I performance is more important. The error floor

varies according to the maximum mobile station velocity. For low velocity
($v = 10$ km/h) it is reached at $C/I \approx 25$ dB whereas for very high velocity
($v = 250$ km/h) it saturates already at $C/I \approx 20$ dB.

A comparison between raw (no channel coding) and coded GSM link performance is shown in Fig. 3.26. The uncoded link curve thus demonstrates the
demodulator performance only (already shown in Fig. 3.23(b)). Coded performance is shown for TCH/F9.6 and TCH/F14.4 traffic channels. Channel

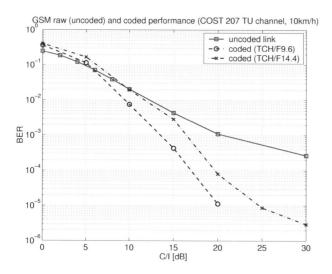

Fig. 3.26. BER of uncoded (raw) GSM link in comparison with the coded link. For
coded link we used TCH/F9.6 and TCH/F14.4. COST 207 Typical Urban channel
model at $v = 10$ km/h was used

coding improves performance for $C/I > 6 - 8$ dB. For smaller C/I the BER
gets worse than for the uncoded link. This is the known effect of every channel
coding scheme. For very bad BER, uncoded links perform better than coded
links. Channel coding works well only if BER is below a specific threshold.
This threshold depends on the used code and decoding algorithm.

The channel coding of TCH/F9.6 gains about two orders of magnitude
in BER at $C/I = 20$ dB, whereas TCH/F14.4 gains only about one order of
magnitude compared to the uncoded link.

An interesting point is also the crossing of the uncoded link performance
with that of the TCH/F14.4 at $C/I \approx 15$ dB. Unless C/I is 15 dB or better,
TCH/F14.4 will behave like an ucoded link. The reason is the less powerful
channel coding.

In Fig. 3.27 we compare the GPRS coding schemes CS-1 to CS-4. For comparison the circuit-switched traffic channels are also included. Both figures,
(a) and (b), are C/I variations, one for $v = 10$ km/h (a) and the right-hand

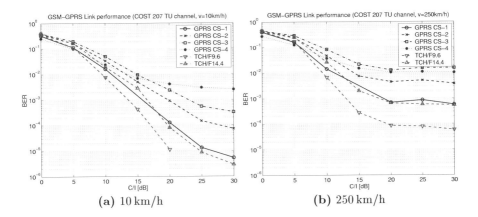

Fig. 3.27. BER of coded GSM link over COST 207 typical urban ('TU') channel as a function of C/I. The mobile station movement was 10 km/h **(a)** and 250 km/h **(b)**. GPRS coding schemes CS-1 to CS-4 together with TCH/F9.6 and TCH/F14.4 traffic channel coding are shown

side one (b) for $v = 250$ km/h. GPRS CS-1, the best GPRS coding scheme has almost the same performance as TCH/F14.4, all other GPRS schemes perform worse. The differences between CS-1 to CS-4 are more pronounced for low mobile station velocity than for high velocity. At low velocities the difference between CS-1 to CS-4 is more than 2 orders of magnitude, from BER $\approx 10^{-5}$ to BER $= 2 \times 10^{-3}$ in the error floor. So the GPRS coding scheme CS-4 still has a very high BER also for $C/I > 25$ dB. Thus the performance of CS-4 in terms of effective available data rate or throughput is expected to be below the nominal one even if the propagation conditions are optimum. Remember that CS-4 has, in essence, no forward error correction at all. This is why CS-4 performs better than CS-3 for $v = 250$ km/h (see Fig. 3.27(b)). For the high velocity cases CS-1 to CS-4 differ about $1\frac{1}{2}$ orders of magnitude of the error floor (from BER $= 6 \times 10^{-4}$ to BER $= 1.5 \times 10^{-2}$). Remarkably, the BER saturates for all GPRS coding schemes at $C/I = 20$ dB for $v = 250$ km/h.

The performance of the GPRS CS-1 scheme for various mobile station velocities is plotted in Fig. 3.28. For low C/I values ($C/I \leq 15$ dB), there is no drop in performance for higher speeds. The difference is more prominent in the error floor. The BER performance for moderate speed up to $v = 50$ km/h is not much worse than for low speed. However, for high speeds the coding schemes differ after all by about $1\frac{1}{2}$ orders of magnitude.

3.3.4 Summary

We presented link level simulations of the physical layer of the GSM radio interface. We implemented the complete transmitter-receiver chain including

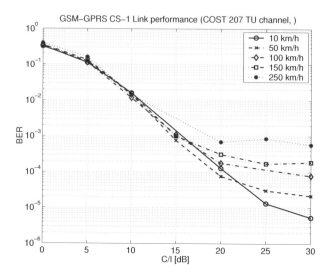

Fig. 3.28. BER of GSM GPRS CS-1 coding scheme over COST 207 typical urban ('TU') channel as a function of C/I. The mobile station movement is the parameter

GMSK modulation, interleaving and forward error correction. We used the COST 207 wideband channel model. This is a tapped delay line model with four different possible area types, 'Typical Urban' (TU), 'Bad Urban' (BU), 'Rural Area' (RA) and 'Hilly Terrain' (HT).

We investigated the BER performance of the uncoded (without FEC) and the full GSM link. Performance of the 'BU' channel model is best with an error floor of only BER $\approx 10^{-5}$ for $v = 10\,\mathrm{km/h}$, almost one order of magnitude below the BER of the 'TU' area type. 'HT' area type yields the worst performance with an error floor of only BER $\approx 4.5 \times 10^{-2}$. A higher velocity (up to $v = 250\,\mathrm{km/h}$) degrades performance at maximum about one and a half orders of magnitude.

BER performance of the full GSM link depends on the channel coding scheme. TCH/F9.6 yields the best performance, gaining about two orders of magnitude compared to the uncoded link at $C/I = 20\,\mathrm{dB}$, wheras TCH/F14.4 gains only one order of magnitude.

The best GPRS coding scheme, CS-1, has almost the same performance as TCH/F14.4. All other GPRS coding schemes are worse, up to two and a half orders of magnitude for CS-4.

3.4 Link Layer Modeling

Link level simulations are computationally *very* expensive tasks. Thus it is desirable to replace link level *simulations* by link layer *models*. Especially for upper-layer protocol considerations a link layer model expands the simulation possibilities extremely by reducing resource requirements. Therefore we will derive and use a model for the complete link layer (physical layer) of the GSM system (i.e. all blocks shown in Fig. 3.17). This model will replace the link level simulations from the previous section. We will use this model in all simulations in Chap. 5. This includes modulation/demodulation, forward error correction and interleaving.

Markov Models are popular for modeling communications channels. Many papers exist dealing with Markov error modeling (e.g. [125,151]). In particular *Hidden Markov Models* (HMMs) are used to describe the bursty nature of communication channels.

A **Hidden Markov Model** (HMM) is characterized by the following properties:

- Number of states, N. The individual states are denoted by $\mathbf{s} = s_1, \ldots, s_N$ and a state at discrete time t as q_t.
- Number of distinct observation symbols per state, M. We denote the possible observation symbols as $\mathbf{v} = v_1, \ldots, v_M$.
- Transition probability distribution, $\mathbf{A} = a_{i,j}[N \times N]$. $a_{i,j}$ is the transition probability from state i to state j,

$$a_{i,j} = P(q_{t+1} = s_j | q_t = s_i), \quad 1 \le i,j \le N . \tag{3.2}$$

- Observation symbol probability distribution, $\mathbf{B} = b_{i,j}[N \times M]$. $b_{i,j}$ is the probability that the symbol v_j is observed as the observation O_t at time t when the HMM is in state q_i,

$$b_{i,j} - P(O_t = v_j | q_t = s_i), \quad 1 \le i \le N; 1 \le j \le M . \tag{3.3}$$

- Initial state distribution, $\boldsymbol{\pi} = \pi_1, \ldots, \pi_N$. π_i is the probability that the initial state is s_i,

$$\pi_i = P(q_1 = s_i) . \tag{3.4}$$

The model is named '*hidden*' because the state sequence can not be determined by observation of the output symbols. A HMM is completely determined by the triple $\lambda = (\mathbf{A}, \mathbf{B}, \boldsymbol{\pi})$ together with the two model parameters N and M. A good introduction into HMMs is given in [112].

HMM parameter estimation from a given observation sequence (e.g. the output of a link level simulation) can be done with the *Baum–Welch-Algorithm* [29] or equivalently with the *Expectation-Modification* (EM) method [44]. Both algorithms are iterative methods, which compute the parameters $(\mathbf{A}, \mathbf{B}, \boldsymbol{\pi})$ by maximizing $P(\mathbf{O}|(\mathbf{A}, \mathbf{B}, \boldsymbol{\pi}))$. We used the EM method for HMM parameter estimation.

Bit vs. block error model. We have to distinguish between *bit error models* and *block error models*. A *block* is a collection of n subsequent bits, where n is the block length. Bit error models generate error sequences for every data bit, whereas block error models only produce an output for every block of bits. A single output sample of the model indicates whether the bit (block) is in error.

In general, a specific HMM can be used to model bit or block errors. Only the error statistic, and thus the HMM parameters, are different. If the bit errors are independent, identically distributed (i.i.d.) the corresponding block error (or Block Error Ratio[3], BLER) is

$$\text{BLER} = 1 - (1 - \text{BER})^n . \tag{3.5}$$

A block is in error if at least one bit within this block is in error. However, this simple transformation is not valid if bit errors have other distributions or are not independent. Especially on systems having burst errors this leads to incorrect block error ratios. So it is important to bear in mind the actual error distribution.

Let us consider the simplest possible HMM, the 2-state model, also known as a Gilbert model [75]. The number of states is fixed, $M = 2$. Output symbols are binary, $M = 2$, '0' for no error and '1' for a bit or block in error. We applied the EM algorithm on an observation sequence computed from link level simulations. The simulations and HMM parameter estimation have been done for each C/I value separately. Initial parameters of the HMM were equally distributed,

$$\boldsymbol{\pi} = [0.5 \ 0.5]; \quad \mathbf{A} = \begin{bmatrix} 0.5 \ 0.5 \\ 0.5 \ 0.5 \end{bmatrix}; \quad \mathbf{B} = \begin{bmatrix} 0.5 \ 0.5 \\ 0.5 \ 0.5 \end{bmatrix}, \tag{3.6}$$

which are inputs into the EM algorithm. The finally estimated HMM parameters for the GSM link level error sequences for specific C/I and velocity values always have the form,

$$\lambda_1 = \left(\boldsymbol{\pi} = [0.5 \ 0.5]; \quad \mathbf{A} = \begin{bmatrix} 0.5 \ 0.5 \\ 0.5 \ 0.5 \end{bmatrix}; \quad \mathbf{B} = \begin{bmatrix} 1 - \text{BLER} & \text{BLER} \\ 1 - \text{BLER} & \text{BLER} \end{bmatrix} \right). \tag{3.7}$$

The estimation process output form is independent of the block length. Of course the detailed values of BLER depend on C/I, v and n. Only the special form of the matrices are the same.

This form of a HMM is equivalent to another popular HMM form. The HMM matrices then have the following form,

$$\lambda_2 = \left(\boldsymbol{\pi}' = [0.5 \ 0.5]; \quad \mathbf{A}' = \begin{bmatrix} 1 - \text{BLER} & \text{BLER} \\ 1 - \text{BLER} & \text{BLER} \end{bmatrix}; \quad \mathbf{B}' = \begin{bmatrix} 1 & 0 \\ 0 & 1 \end{bmatrix} \right). \tag{3.8}$$

See also Fig. 3.29. In HMM Form 2 there exists one state, the 'good' state,

---

[3] We use the term 'Block Error Ratio', BLER, to distinguish it clearly from the 'Bit Error Ratio', BER.

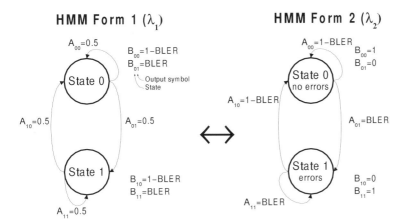

Fig. 3.29. Two-state Markov model with parameters estimated from simulated GSM link level simulations (*left-hand side*). An equivalent HMM is shown on the *right-hand side*. The statistical output symbol properties are the same for both models

where no errors occur and a second state, the 'bad' state, where an error always occurs. This Form is shown on the right-hand side of Fig. 3.29.

Two HMMs, λ_1 and λ_2, are equivalent, if they have the same statistical properties for the observation symbols, i.e. $E[O_t = v_k|\lambda_1] = E[O_t = v_k|\lambda_2]$ for all symbols v_k [112]. In the special case of a 2-state model this leads to

$$A_{0,0}^1 B_{0,0}^1 + A_{1,0}^1 B_{1,0}^1 = A_{0,0}^2 B_{0,0}^2 + A_{1,0}^2 B_{1,0}^2 \,, \tag{3.9}$$

$$A_{0,1}^1 B_{0,1}^1 + A_{1,1}^1 B_{1,1}^1 = A_{0,1}^2 B_{0,1}^2 + A_{1,1}^2 B_{1,1}^2 \,, \tag{3.10}$$

where the superscript indicates the model. If (3.9) and (3.10) are fulfilled then the two models, λ_1 and λ_2, are equivalent with respect to their statistical properties of the output symbols. Equations (3.9) and (3.10) are satisfied for the models given in (3.7) and (3.8). Thus the two forms are proven to be equivalent.

Looking at the structure of the first model, λ_1, we can observe that the two states are completely equivalent. The observation probabilities for each symbol are the same in every state. Also the probability of being in a specific state is *equal* for both states. This fact lets us conclude that this model produces an i.i.d. error pattern. Thus the GSM link level including interleaving and channel coding also produces an error pattern which is independent and identically distributed. So after channel coding and interleaving, *no* burst errors exist any longer. All errors are independent of each other. Remember that there is *no* C/I variation included.

The i.i.d. property of the output error sequence can be verified by computing the autocorrelation of this sequence. Figure 3.30 shows the autocorrelation, Rxx, for $C/I = 10\,\text{dB}$ and $v = 10\,\text{km/h}$ (a) and $C/I = 5\,\text{dB}$ and

$v = 250\,\mathrm{km/h}$ (b) for the GPRS CS-1 coding scheme. The output error sequences are obtained from the link level simulations described in the previous section. The block length was set to $n = 184$, which is the RLC/MAC

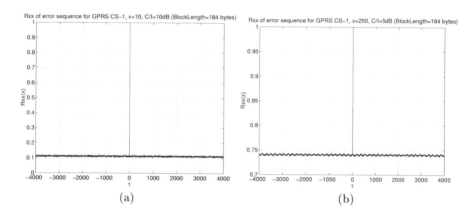

Fig. 3.30. Autocorrelation of error sequence from GPRS CS-1 channel coding with $C/I = 10\,\mathrm{dB}$, $v = 10\,\mathrm{km/h}$ (a) and $C/I = 5\,\mathrm{dB}$, $v = 250\,\mathrm{km/h}$ (b). The block length was fixed at $n = 184$.

block length in GPRS CS-1. GPRS coding schemes have the least interleaving within the GSM system over just four bursts. Thus the circuit-switched data channels, which interleave over 19 bursts, have even more independent output bits (for a specific C/I). The Dirac-shaped autocorrelation functions indicate that error bits are almost fully uncorrelated. Thus the i.i.d. property is corroborated, and the simple 2-state model is sufficient to approximate an error sequence. This is also shown in [150]. However, we always have to remember that this result is only valid for a *single* specific C/I value. No C/I variation was included here.

C/I variation with BER/BLER. The dependence of the BER on the C/I value has already been evaluated in the link level simulations (e.g. Fig. 3.27 and Fig. 3.28). In this section we present a model for the BER and BLER dependence on the C/I. With such a model the BER (or BLER) value necessary for the Markov model can be easily determined for a given C/I parameter. Here C/I is taken as an input parameter.

One way to pack a BER-curve into a suitable form for simulation would be *lookup tables*. This is a standard method (see for example [147]). It requires a number of points to obtain good interpolation results between the stored base points. Otherwise modeling for intermediate values will be inaccurate.

Another way is to fit a function to the BER(C/I) or BLER(C/I) function. Looking at the shape of the function (see sketch in Fig. 3.31), we considered a low-pass filter function suitable. In the following we will use only the term block error ratio, BLER, which depends on the block length, n. The bit error ratio, BER, is included therein with $n = 1$. The BLER can be approximated according to a low-pass filter function by

$$\text{BLER} = \frac{\text{SV} - \text{EF}}{1 + ((C/I)/\text{CIc})^k} + \text{EF} ,\tag{3.11}$$

where SV (Starting Value) is the BLER for very low C/I, EF (Error Floor) the BLER for very high C/I and CIc denotes the *cut-off* C/I value where the BLER is 3 dB below SV. k determines the slope of the function between SV and EF. That is, the model is fully characterized by four parameters, SV, EV, CIc and k (Fig. 3.31). The slope parameter, k, can be obtained from

$$k = \frac{\log(\frac{\text{BLER1}-\text{SV}}{\text{EF}-\text{BLER1}})}{\log(\text{CI1}/\text{CIc})} .\tag{3.12}$$

Here BLER1 is the BLER at $C/I = \text{CI1}$; i.e. $\text{BLER1} = \text{BLER}(CI1)$. The pair $(\text{BLER1}, \text{CI1})$ must be in the slope section of the function.

The model parameters, SV, EF, CIc, themselves depend on system parameters (used traffic channel), channel parameters (velocity and propagation class) and the block length,

$$\text{SV} = f(n), \quad \text{EF} = f(\text{TCH}, v, \text{PropClass}, n),$$
$$\text{CIc} = f(\text{TCH}, v, \text{PropClass}, n),\tag{3.13}$$

where PropClass is one of 'Typical Urban', 'Bad Urban', 'Hilly Terrain' or 'Rural Area'. The starting value, SV, only depends on the block length, n,

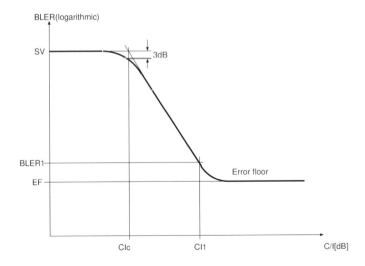

Fig. 3.31. Sketch of a BER (BLER) simulation with C/I as parameter. Also included are parameters for modeling this curve

because the BER for very low C/I is always 0.5. Only a concatenation of bits to a block could increase this value. In the limiting case of a block with infinite length the BLER is 1, thus $SV(\infty) = 1$.

Link level simulation results are the source for the model parameter extraction. We now turn to extracting specific parameter values for GPRS CS-1 to CS-4 channel coding scheme and 'Typical Urban' propagation class from the link level simulations (see previous section). The parameters are summarized in Table 3.9. The starting value, SV, as well as the cut-off C/I value, CIc remain constant for a specific block length. Also the slope parameter, k, does not show much variation for different velocities. The dependence of the error floor, EV, on the velocity, v, and the block length, n, is illustrated in Fig. 3.32. The used coding scheme is GPRS CS-1 and 'Typical Urban' prop-

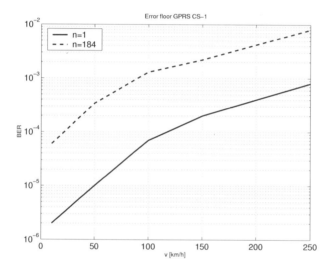

Fig. 3.32. Dependence of the error floor, EF, on the velocity of the mobile station for GPRS CS-1 coding scheme and 'TU' propagation class. The block length parameter is $n = 1$ and $n = 184$

agation model was used. The error floor increases with the velocity of the mobile station. The rise is sharper for lower speeds and it starts to saturate for high velocities. For a block length of $n = 184$ the error floor is approximately one order of magnitude above the one for $n = 1$ (which equals the BER case).

An evaluation of (3.11) for GPRS CS-1, 'TU' and velocities of $v = 10\,\text{km/h}$ and $v = 250\,\text{km/h}$ is plotted in Fig. 3.33. The block length was fixed to $n = 1$, thus the BER is modeled. Compared to the original simulation result in Fig. 3.28, the model shows adequate conformance with simulated data.

Table 3.9. Model parameters for GPRS CS-1 (**a**), CS-2 (**b**), CS-3 (**c**) and CS-4 (**d**) coding scheme, 'Typical Urban' propagation class and various velocities. The block length is $n = 1$ (i.e. BER) and $n = 184$ (**a**), $n = 271$ (**b**), $n = 315$ (**c**) and $n = 431$ (**d**), which equals the GPRS RLC/MAC block size in the according coding scheme

(**a**) GPRS CS-1

| v | | $n = 1$ | | | | $n = 184$ | | |
(km/h)	SV	EF	CIc	k	SV	EF	CIc	k
10	0.5	2×10^{-6}	1 dB	3.74	1	6×10^{-5}	5 dB	3.84
50	0.5	2×10^{-5}	1 dB	4.01	1	2×10^{-4}	5 dB	3.77
100	0.5	7.8×10^{-5}	1 dB	3.90	1	1×10^{-3}	5 dB	3.69
150	0.5	1.8×10^{-4}	1 dB	3.98	1	2×10^{-3}	5 dB	3.69
250	0.5	5.5×10^{-4}	1 dB	4.12	1	8.5×10^{-3}	5 dB	3.94

(**b**) GPRS CS-2

| v | | $n = 1$ | | | | $n = 271$ | | |
(km/h)	SV	EF	CIc	k	SV	EF	CIc	k
10	0.5	5×10^{-5}	1.5 dB	3.03	1	1.1×10^{-3}	6 dB	2.79
50	0.5	2×10^{-4}	1.5 dB	3.13	1	3×10^{-3}	6 dB	2.65
100	0.5	8×10^{-4}	1.5 dB	3.07	1	1.3×10^{-2}	6 dB	2.69
150	0.5	1.3×10^{-3}	1.5 dB	3.13	1	2.3×10^{-2}	6 dB	2.55
250	0.5	3.6×10^{-3}	1.5 dB	3.01	1	9×10^{-2}	6 dB	2.56

(**c**) GPRS CS-3

| v | | $n = 1$ | | | | $n = 315$ | | |
(km/h)	SV	EF	CIc	k	SV	EF	CIc	k
10	0.5	3×10^{-4}	2.2 dB	2.72	1	4×10^{-3}	6 dB	2.16
50	0.5	9×10^{-4}	2.2 dB	2.93	1	1.2×10^{-2}	6 dB	2.24
100	0.5	1.8×10^{-3}	2.2 dB	2.70	1	3.2×10^{-2}	6 dB	2.00
150	0.5	3.2×10^{-3}	2.2 dB	2.72	1	6.6×10^{-2}	6 dB	2.11
250	0.5	1.3×10^{-2}	2.2 dB	2.78	1	4×10^{-3}	6 dB	1.54

(**d**) GPRS CS-4

| v | | $n = 1$ | | | | $n = 431$ | | |
(km/h)	SV	EF	CIc	k	SV	EF	CIc	k
10	0.5	2.5×10^{-3}	-1 dB	2.61	1	7.5×10^{-1}	12 dB	2.73
50	0.5	2.6×10^{-3}	-1 dB	2.61	1	7.6×10^{-1}	12 dB	2.87
100	0.5	4×10^{-3}	-1 dB	2.55	1	7.8×10^{-1}	12 dB	2.84
150	0.5	5.2×10^{-3}	-1 dB	2.57	1	8.25×10^{-1}	12 dB	3.07
250	0.5	1×10^{-2}	-1 dB	2.61	1	8.75×10^{-1}	12 dB	2.18

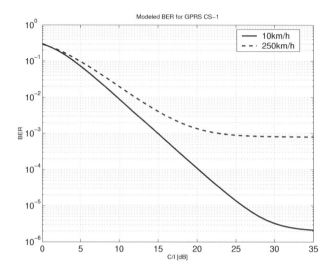

Fig. 3.33. Evaluation of (3.11) for GPRS CS-1 coding scheme, 'TU' propagation class and two different velocities ($v = 10\,\mathrm{km/h}$ and $v = 250\,\mathrm{km/h}$). The block length is $n = 1$, therefore BER is shown. The model corresponds well to the simulated BER performance (Fig. 3.28)

The evaluation of (3.11) for GPRS CS-1, CS-2, CS-3 and CS-4 with the block error parameters in Table 3.9(a–d), right-hand side, is plotted in Fig. 3.34. These are the final block error ratios, which we used in further GPRS evaluations (Chap. 5).

Variations in the input C/I according to a specific statistical distribution can also be considered very easily now. The instantaneous C/I is generated according the statistical properties and the corresponding BLER can be computed by (3.11).

With this model (3.11) and a set of model parameters for various system and channel parameters it is possible to replace the link level simulation with a time-saving but nevertheless sufficiently accurate method. We will use this model in further evaluations of upper layer protocols and system parameters throughout this book.

3.5 Summary

In GSM Phase 2+ there exist circuit-switched (Circuit-Switched Data, CSD) and packet-switched bearer services. The most important circuit-switched services are TCH/F9.6 (Traffic Channel, Full rate at 9.6 kbit/s user data rate) and TCH/F14.4 (14.4 kbit/s). In multislot operation (High-Speed Circuit-

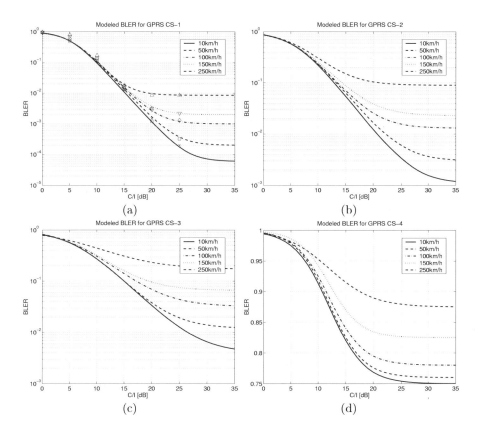

Fig. 3.34. Evaluation of (3.11) for GPRS CS-1 (**a**), CS-2 (**b**), CS-3 (**c**) and CS-4 (**d**) coding scheme, 'TU' propagation class and various velocities. The block error ratio is shown. For CS-1 (**a**) also the simulated data points are plotted

Switched Data, HSCSD) n timeslots per frame can be used for a traffic channel, thus increasing the data rate by a factor of n compared to TCH/F9.6 or TCH/F14.4. However, the maximum A-interface data rate of 64 kbit/s limits the HSCSD data rate. Therefore using all 8 timeslots simultaneously is *not* possible in practice.

The General Packet Radio Service, GPRS, provides packet-switched bearer services within GSM. The entire transmission line from mobile station, radio interface, GPRS core network to the wired public packet-switched data network is packet switched. Scarce radio resources are only allocated when there is data to send and a multiplexing gain is achieved between several users on the radio interface. Four new channel coding schemes, CS-1 to CS-4, are introduced, which differ in the net user data rate, i.e. the protection quality of the coding is different. Thus, and also because of multislot operation on the radio interface, the maximum user data rate is about 170 kbit/s. But this maximum rate is obtained only at very good propagation conditions for a single user and no additional network load. We will show quantitative values in Chap. 5. In practice under real network conditions the available data rate will be only a fraction of the maximum one.

In both circuit-switched and packet-switched modes there exist transparent and non-transparent services. Transparent service provides constant throughput (the nominal throughput) where only forward error correction is applied. In non-transparent operation an additional ARQ protocol at layer 2 (Radio Link Protocol, RLP, in CSD and Radio Link Control Protocol, RLC, in GPRS) protects the data stream. Due to possible retransmissions the data rate is not constant but depends on actual transmission quality. But the BER is lower than in transparent mode, about BER $\leq 10^{-9}$. Although the layer 2 ARQ protocols are, in principle, the same for CSD and packet-switched data there exist one major difference: In CSD the RLP spans from mobile station to Mobile-services Switching Center, MSC, and in GPRS the RLC protocols spans only from mobile station to the Base Station Subsystem, BSS.

We evaluated the performance in terms of BER of the GSM radio link for both CSD and GPRS by means of link level simulations. COST 207 wideband channel model is the applied model for the radio environment. Simulation includes the complete GSM physical layer. Results are available for TCH/F9.6 and TCH/F14.4 CSD traffic channels as well as for all 4 GPRS coding schemes CS-1 to CS-4. Because today's GSM systems are usually interference limited, BER as a function of the carrier-to-interference ratio, C/I, is the most important performance measure. TCH/F9.6 has the best performance and yields BER $= 10^{-5}$ at $C/I = 10\,$dB for a mobile station speed of $v = 10\,$km/h and the typical urban (TU) channel model. This is almost one order of magnitude better than TCH/F14.4. The best GPRS coding scheme, CS-1, achieves the same performance as TCH/F14.4. All other GPRS coding schemes, CS-2 to CS-4 are worse. The range is about $1\frac{1}{2}$ ($v = 250\,$km/h) up to 2 ($v = 10\,$km/h) orders of magnitude. BER performance evaluated with the bad urban (BU) channel model is about one order of magnitude below that for TU.

We used a 2-state Hidden Markov Model (HMM) as replacement for the complete GSM link level including FEC, modulation/demodulation and channel model. For extraction of the model parameters we used the Expectation-Modification method applied to simulated data. It turns out that the error sequences are i.i.d. distributed for a specific C/I value. The model parameters depend on the traffic channel, mobile station velocity, and radio channel propagation class. For modeling block errors the same model as that used for bit errors can be applied with different parameters. The dependence of BER (BLER) on C/I can be expressed with a simple 'low pass filter' function. The parameters of the function are obtained by curve fitting. This leads to a simple link level model which enables extensive and time-saving evaluation of higher-level protocols.

4. Data Transmission in UMTS

The *Universal Mobile Telecommunication System* (UMTS) is the third-generation mobile radio system that will prevail in Europe and Japan. It is currently under standardization, which is done by *3GPP*, 3rd Generation Partnership Project. 3GPP is a partnership of several standardization bodies around the world. As the specifications are still under development, all subsequent elaborations refer to the *Release'99* of the standard, published in March 2000. Exceptions are explicitly mentioned. The technical specifications are publicly available on the Internet at http://www.3gpp.org/. As for GSM in the previous chapter, we will restrict ourselves here with those parts that are relevant for this book.

Like GSM, UMTS is a hierarchical system. An overview of a possible UMTS system is outlined in Fig. 4.1. The thick paths show the data path interfaces where user data is transmitted. Only these interfaces are shown in Fig. 4.1. Control interfaces as well as location, subscriber and network management entities are omitted for clarity.

UMTS has two main parts: the *Core Network* (CN) and the *Access Network* (AN) [10].

4.1 Core Network (CN)

In Release'99 an evolved GSM core network is specified. However, ongoing studies investigate other technologies for a future UMTS core network as well. Thus we provide also a brief overview for an *ATM-based* core network and an *All IP* core network.

4.1.1 Release'99 Core Network

The Release'99 core network follows an evolutionary path from GSM and uses (enhanced) main parts of the GSM core network [10]. It is depicted in the upper part of Fig. 4.1. In essence, the core network is the same as in GSM. All protocols and interfaces *within* the core network have been already specified in GSM Phase 2+. But to support higher data rates upgrading of all entities involved is necessary. The Release'99 core network supports both

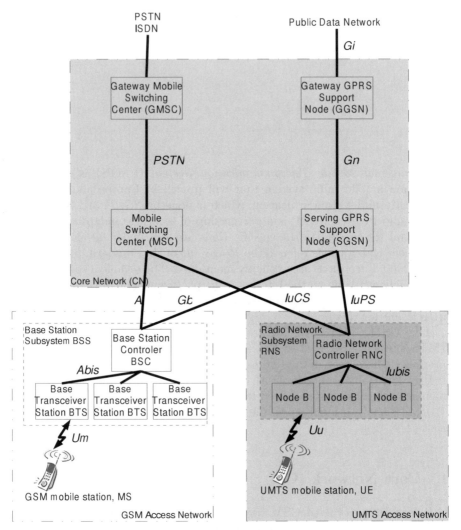

Fig. 4.1. Overview of a UMTS network with UTRA and GSM type radio access networks [10]

the conventional GSM radio access network, via *A* and *Gb* interfaces, as well as the new UMTS Terrestrial Radio Access Network (UTRAN). UTRAN is connected to the CN via *IuCS* for circuit-switched services and *IuPS* for packet-switched services (corresponding to *A* and *Gb* in GSM). It should be possible to operate a GSM (and GSM GPRS) access network in parallel with the new UMTS access network. Both will be connected to a common core network (Fig. 4.1).

The CN is logically divided into the *circuit-switched domain* (CS) and the *packet-switched domain* (PS). Entities common to PS and CS domains are location registers, authentication center, equipment identity register and Short Message Service (SMS) related entities. These entities are omitted in Fig. 4.1 for clarity.

The circuit-switched domain includes Mobile-services Switching Center (MSC), Gateway-MSC and the Interworking Function (IWF). They have the same principal functionality as in GSM. All circuit-switched connections are passed through the CS domain. These are speech and circuit-switched data connections.

The packet-switched domain is evolved from the GSM-GPRS core network. It includes the Serving GPRS Support Node (SGSN) and the Gateway GPRS support node (GGSN). All packet-switched services are done in this domain. The PS domain is in parallel to the CS domain.

The **protocol architecture** in the core network is the same as in GPRS core network. The *Gn* interface between SGSN and GGSN uses the GPRS Tunneling Protocol (GTP) for transmission of user data. Below the GTP a TCP/IP or UDP/IP stack is used. See Sect. 3.2.4 and Sect. 4.3 for further details.

4.1.2 ATM-based Core Network

The *Asynchronous Transfer Mode* (ATM) is currently being investigated as a candidate transport technology in a future UMTS core network [2]. UMTS core network requirements include support for circuit-switched and packet-switched traffic, different QoS requirements and support for real-time, non-real-time and adaptive flow control services.

ATM provides a connection-oriented approach which makes it easy to support stringent QoS requirements. These are necessary, in particular for real-time services (speech) as well as data connections with specific QoS requirements. The authors of [2] conclude that the ATM statistical multiplexing gain[1] is larger than that of an IP packet network. Also the required buffer size for data connections is lower for ATM with AAL2 (ATM Adaptation layer type 2). For example, at a link utilization of 95% a buffer size of 20 000 bytes is needed in an ATM network, whereas 33 000 bytes are needed for an IP network. Furthermore, ATM supports more simultaneous voice connections when used with AAL type 2 than an UDP/IP network.

Thus ATM seems to be a suitable core network transport technology. As a further benefit, ATM networks are widely used and ATM is scalable.

The basic protocol between SGSN and GGSN remains GTP (GPRS Tunneling Protocol) which tunnels the user IP packets between the GSNs. Below

---

[1] In packet-switched networks, packets are not sent during the entire connection time. Also, the packets are sent in different time intervals. If several traffic flows are multiplexed over a single connection a *statistical multiplexing gain* is achieved due to idle times in the data flows. These idle times can be filled by other traffic.

GTP TCP/IP or UDP/IP is used for addressing of the GSNs (see Fig. 4.2). However, the IP protocol below GTP is not used for routing but *only* for addressing. The ATM sublayer is responsible for routing between the GSNs. A transport layer converter converts the IP addresses into ATM addresses.

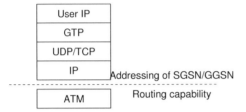

Fig. 4.2. User plane protocol stack for a packet core network between GSNs [2]

4.1.3 All IP Network

The *All IP network* is based on the evolution of GPRS [1]. Release'00 should contain an All IP network. Strictly speaking, the All IP network also covers the radio access network. The architecture is packet switched. It should support voice, data, real-time multimedia and services with the same network elements. Also circuit-switched services should be supported by this packet-switched network. The main benefits of using this approach are: Seamless service (always IP regardless of the service type), synergy with IP developments and reduced cost of service, efficient solution for simultaneous services including voice and data, direct support of Internet applications, cost reduction through packet transport.

Two reference architecture options are currently under consideration [1]:

- Option 1 supports an IP-based architecture *only* based on packet technologies and IP telephony for simultaneous real-time and non-real-time services. It includes the radio network, the GPRS network, and a completely new call control and service architecture. The Signalling System 7 (SS7) is replaced by IP. The network architecture is independent of layer 1 and layer 2. Also, a separation of service control from call/connection control is provided.
- Option 2 supports *additionally* Release'99 circuit-switched terminals. The support can be done in two ways: Conventional MSCs and GMSCs are involved for routing the call. Or a *Media Gateway Function* is used to interwork the circuit-switched call at the *Iu* interface into real-time packet-switched services.

Both options introduce a number of new functional elements for call control, location management, and signalling gateways to other networks.

Transmitting voice in IP packets (Voice over IP, VoIP) requires special consideration for the radio interface. Optimal coding and interleaving must be applied for this type of packet. Radio link optimization for real-time IP includes header compression of IP/UDP/RTP[2] header which carries small voice payload packets. This greatly improves overhead efficiency. Particular considerations have to be made for the case of packet loss. Header compression is usually sensitive to packet loss, because only differential information of the headers is stored in the compressed header. Especially for real-time services, no acknowledged data transfer mode can be used to guarantee the time constraints. Thus header compression may not be as optimal as in the fixed network case. Another approach is to provide stronger error protection for the compressed header. This has to be considered also in the light of the required service quality. Three different service categories are considered: Basic voice, where payload optimization (header stripping or compression) together with unequal error protection of header and data is used. Real-time multimedia services, here the detailed parameters are not yet defined. And third is the Pure IP service, where the bearer service does *no* adaptation of the data stream.

4.2 Access Network, UMTS Terrestrial Radio Access Network (UTRAN)

The UMTS access network is called *UMTS Terrestrial Radio Access Network* (UTRAN). The UTRAN is specified in the 3GTS25.xxx series of the 3GPP specifications. UTRAN introduces a completely new radio interface, which should provide data rates up to 2 Mbit/s. An overall description can be found in [19].

UTRA architecture consists of *Radio Network Controllers* (RNC) and *Node B*s. One RNC and the connected Node Bs form a *Radio Network Subsystem* (RNS). See Fig. 4.1 for illustration. More than one RNS can be connected to the core network. A RNC is linked via the *Iu* interface to the core network (*IuCS* and *IuPS*). The *User Equipment* (UE) is connected to a Node B via the *Uu* interface. UE is a new synonym in UMTS for the mobile station.

Every UE has a connection to a RNS, i.e. to a RNC via a Node B controlled by this specific RNC, called serving RNC in the serving RNS. If required, a second RNC can support additional or stand-alone connection to the UE. This second RNS (RNC) is called *drift* RNS (drift RNC). This scenario is likely during an inter-RNS handover as shown in Fig. 4.3. A special SRNS relocation function is used to make the drift RNS to the new serving RNS (SRNC Relocation). If two RNCs are involved in a single connection it has impact especially on the Medium Access Control (MAC) protocol layer (see Sect. 4.3.2).

---

[2] Real Time Protocol, RTP

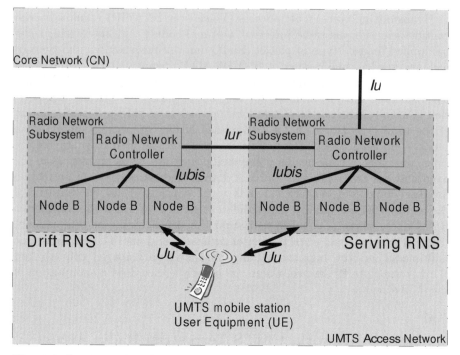

Fig. 4.3. Serving and Drift Radio Network Subsystem (RNS) [19]

Radio interface. The radio interface is based on *Direct Sequence Code Division Multiple Access* (DS-CDMA) multiple access technology. Two duplex modes exist: Frequency Division Duplex (FDD) and Time Division Duplex (TDD) [13].

In UTRA TDD there is a TDMA component in addition to DS-CDMA. Thus it is often denoted as TD/CDMA. A physical channel in FDD is characterized by the code, frequency and, in the uplink, the relative phase (I/Q). In the TDD mode the physical channels are additionally determined by the timeslot.

A 10-ms radio frame is divided into 15 timeslots. The CDMA chip rate is 3.84 Mcps which leads to 2560 chips per slot. The information rate varies with the symbol rate that is determined by the chip rate divided by the spreading factor (SF). The 10-ms frame is the basic unit for data transmission and for higher-layer protocols. The data transport services offered by the physical layer to higher layers operate according the radio frame timing. Every transmission block is generated precisely every 10 ms or a multiple of 10 ms. A *Transport Block* is defined as the data accepted from the physical layer which is jointly encoded. It is the basic unit of data exchange between physical layer and MAC layer. Typically a transport block corresponds to a

Radio Link Control Protocol Data Unit (RLC PDU). The physical layer adds error control coding and CRC.

Transport Block Sets are a set of transport blocks which are exchanged at the same time between physical and MAC layer using the same transport channel [17]. The inter-arrival time of transport block sets is called the *Transmission Time Interval* (TTI). It is always a multiple of the minimum interleaving period, which is also 10 ms. Currently TTIs of 10, 20, 40 and 80 ms are defined. See Fig. 4.4 for illustration. The *Transport Block Format* (TBF)

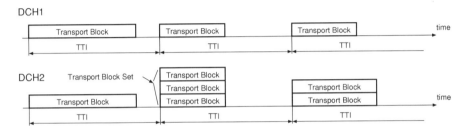

Fig. 4.4. Exchange of Transport Blocks and Transport Block Sets between physical and MAC layer via two parallel Dedicated Channels (DCH1 and DCH2) [17]

determines the format of the transport block. It consists of a *dynamic* and a *semi-static* part. The dynamic part determines *Transport Block Size*, *Transport Block Set Size* and transmission time interval (only in TDD, optional a dynamic attribute). The semi-static part determines the transmission time interval (mandatory for FDD, optional dynamic for TDD bearers), and the applied error protection scheme. This includes the type of error protection (convolutional code, turbo code or no channel coding), coding rate, static rate matching parameter, puncturing limit and CRC size. Transport formats are combined to *Transport Format Sets* associated to a transport channel. The semi-static parts stay the same within a transport format set. The dynamic part may vary within a transport format set for different transport formats. It is the semi-static part that determines the service attributes like transmission quality (e.g. BER) and transfer delay.

Channel coding and **multiplexing** in UMTS depends on the requested bearer capability. It is defined in [8] for FDD and [9] for TDD. It is a multistage process. First a CRC is added to each transport block. Then all transport blocks corresponding to one transport block set are concatenated and afterwards segmented into *Code Blocks*. Each code block is encoded separately and all encoded blocks are concatenated again. This encoded concatenated blocks are interleaved (1st interleaving) and hereafter segmented into radio frames and rate matched. This was coding for a single Transport Channel. The radio frames of several transport channels are multiplexed to *Coded Com-*

posite Transport Channels (CCTrCH). If more than one physical channel is used, the *physical channel segmentation* divides the CCTrCH into the different physical channels. Then 2nd interleaving is performed separately for each physical channel. Both interleavers are block interleaver with inter-column permutations.

For lower data rates convolutional coding with rate 1/2 or 1/3 is applied. Higher data rates use a parallel turbo coding scheme with trellis termination and a rectangular interleaver with intra-row and intra-column permutations [8,9].

An example for channel coding for a 64/128/384-kbit/s packet data channel on a Dedicated Channel (DCH) in FDD uplink and a 64-kbit/s data service also in FDD uplink is shown in Fig. 4.5(a) and (b), respectively. The parameters of the examples are summarized in Table 4.1.

Table 4.1. Parameters for the channel coding example in Fig. 4.5 [3]

Parameter		64/128/384 kbit/s packet data	64 kbit/s data circuit switched
Number of TrChs		1	1
Transport Block Size		640 bits	640 bits
Transport	64 kbit/s	640×B bits (B=0,1)	4×640 bits
Block	128 kbit/s	640×B bits (B=0,1,2)	–
Set Size	384 kbit/s	640×B bits (B=0,1,2,...,6)	–
CRC		16 bits	16 bits
Coding		Turbo coding, code rate=1/3	
Transmission Time Interval (TTI)		10 ms	40 ms

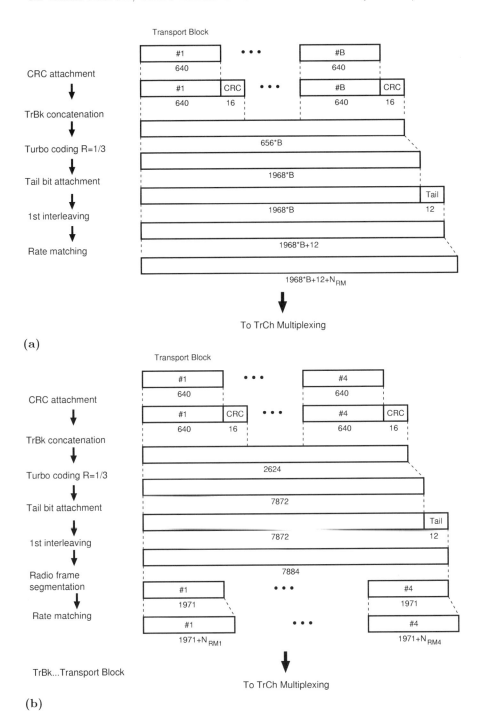

Fig. 4.5. Channel coding examples for a 64/128/384-kbit/s packet data channel **(a)** and a 64-kbit/s circuit-switched data channel **(b)**. Both examples belong to FDD mode in uplink. The corresponding parameters are listed in Table 4.1 [3]

4.3 Protocol Architecture

The entire protocol architecture is divided into a *control plane* and a *user plane*. Actual data transfer is done within the user plane. We will describe only the user plane protocols in the next sections. Also, we will restrict this overview to packet-switched data transmission modes. Circuit-switched data bearers use the conventional GSM system structure in the core network and thus also use those protocols.

The structure of the user plane and control plane protocols for packet switched data transmission is given in [6].

4.3.1 Overall Protocol Architecture (User Plane)

The overall UMTS protocol architecture for the user plane is drawn in Fig. 4.6 [6]. It is similar to the GSM GPRS architecture, especially in the core net-

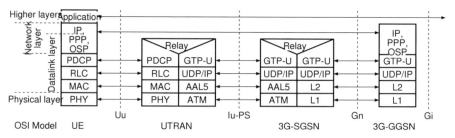

Fig. 4.6. Overall UMTS protocol architecture for the packet-switched domain and user plane [6]

work. The radio interface protocol architecture (UTRAN) is explained in detail in the next section. To differentiate SGSN and GGSN for GSM-GPRS and UMTS, the 3rd generation GSNs are called 3G-SGSN and 3G-GGSN.

Iu-interface
The *Iu*-interface user plane protocols are defined in [18]. In the packet-switched user plane domain the *GPRS Tunneling Protocol* for the user plane (GTP-U) is applied. As transport layer below GTP the User Datagram Protocol (UDP) over IP is applied. IPv6 support is optional. Classical IP over ATM protocol encapsulation over ATM Adaption Layer 5 (AAL5) will be used to carry the IP packets over the ATM transport network ([89]).

Gn-interface
Almost the same protocol stack as for *IuPS* interface is defined for the *Gn*-interface in the user plane. Again GTP-U tunnels the network layer packets from 3G-SGSN to 3G-GGSN. The transport tunnel is UDP/IP. The low level transport system is not specified explicitly. In principle any type of transport technology (ATM, IP-based, ...) can be used (see also Sect. 4.1).

4.3.2 Radio Interface Protocol Architecture

The radio interface protocol architecture defines protocols spanning the *Uu* interface between User Equipment and Node B and also to the Radio Network Controller (RNC). It is specified in [15]. It is divided into three protocol layers, the physical layer (L1), the data link layer (L2) and the network layer (L3). Layer 3 services exist only in the control plane.

Figure 4.7 shows Layer 1 and Layer 2 protocols. They are divided into

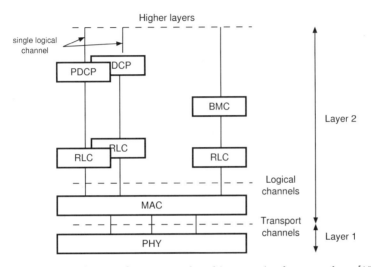

Fig. 4.7. Radio interface protocol architecture in the user plane [15]

physical layer, Medium Access Control (MAC) layer, Radio Link Control (RLC) layer, Packet Data Convergence Protocol (PDCP) layer and Broadcast/Multicast Control (BMC) layer. A vertical line in Fig. 4.7 represents a logical channel. The MAC layer is responsible for mapping of the logical channels to the transport channels. It is shared for all logical channels. The higher (sub-)layer protocols (RLC, PDCP and BMC) have separate entities for every logical channel. The Broadcast/Multicast Control (BMC) exists only for broadcast/multicast logical channels. We will not further describe BMC. Details of MAC, RLC and PDCP layers will follow in the next sections.

Medium Access Control, MAC. The Medium Access Control ([7]) consists of three different functional entities:

- **MAC-b**: This handles the Broadcast channel (BCH). There exists one entity in the UE and one in the UTRAN for every cell.
- **MAC-c/sh**: This entity handles all common and shared channels, like paging, access, common packet and shared data channels. One MAC-c/sh is located in the UE and one in the UTRAN for every cell.

The following functions are performed:

The scheduling and priority module within MAC-c/sh manages the resources between different UEs and different data flows. The Transport Format Combination (TFC) used in the physical layer is selected in the MAC layer. This TFC can be one from the active Transport Format Combination Set determined from the Radio Resource Control layer (RRC). RRC is located in the control plane of Layer 3. MAC-c/sh also manages the flow control of the MAC-d entities. If controlling and serving RNC are not identical, MAC-s/sh is located in the controlling RNC (see also Fig. 4.12).

- **MAC-d**: MAC-d controls all dedicated channels, including logical and transport channels. A MAC-d entity is in the UE and in the UTRAN for *every* UE that is camped in the cell.

The most important functions of MAC-d are outlined below:

- Dynamic transport channel type switching based on decisions of the RRC.
- Multiplexing of several dedicated logical channels onto one transport channel. Higher layer PDUs are multiplexed into transport blocks determined for transport channels.
- Ciphering/deciphering for transparent RLC operation in MAC layer. In transparent RLC operation (see below) ciphering can be done in the MAC layer.

The MAC data PDU consists of an optional MAC header and a MAC Service Data Unit (SDU). Both have variable size. The MAC header may consist of several fields. Possible combinations of different MAC header field entries are shown in Fig. 4.8. The combinations are valid for Dedicated Control Channel (DCCH) and Dedicated Traffic Channel (DTCH). For other channels, other combinations and sizes are defined. If a dedicated channel is mapped onto a forward access channel (FACH) or random access channel (RACH) then the *Target Channel Type Field* (TCTF) provides the identifica-

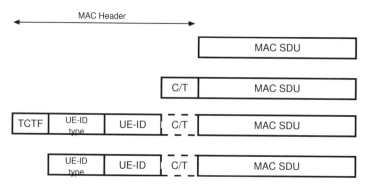

Fig. 4.8. Possible combinations for a MAC PDU for a Dedicated Traffic Channel (DTCH) or Dedicated Control Channel (DCCH) [7]

tion of the dedicated channel. For mapping onto FACH-FDD, the TCTF field is 5 bits (01100), for mapping onto RACH-FDD, the TCTF field is 2 bits. In TDD mode TCTF field is either 1 bit (1) when mapped onto a shared channel or 4 bits (0100) when mapped onto RACH-TDD.

The *C/T field* provides identification of the logical channel instance if multiple logical channels are multiplexed on the same transport channel. If no multiplexing is done, C/T field is not required. The length of the C/T field is 4 bits, thus leading to a maximum of 15 multiplexed logical channels in a single transport channel (one 4-bit word is reserved).

If dedicated channels (DTCH or DCCH) are mapped onto RACH, FACH or a shared channel, the *UE-ID field* and the *UE-ID Type field* are mandatory. UE-ID Type specifies the type of the UE-ID field, it is 2 bits wide. UE-ID field is either 16 bits wide (for DTCH) or 32 bits wide (for DCCH).

Table 4.2 clarifies the use of the MAC header fields for different DTCH and DCCH channel mappings.

Table 4.2. Possible MAC header fields for DCCH and DTCH channel for mapping on various transport channels

DTCH or DCCH mapped on...	TCTF	MAC header field		
		UE ID type	UE ID	C/T
FACH-TDD	5 bits (01100)	16		0
RACH-TDD	2 bits	or	2 bits	or
shared TDD	1 bit	32 bits		4 bits
RACH-TDD	4 bits (0100)			
DCH [a]		No MAC header		
DCH [b]	-	-	-	4 bits

[a] With no multiplexing of dedicated channels on MAC (i.e. a single DTCH or DCCH on a single DCH).
[b] With multiplexing of dedicated channels on MAC (i.e. multiplexing of several DTCH or DCCH on a single DCH).

The MAC layer is also responsible for measurements reported to RRC layer. For example, these are traffic volume monitoring functions.

Radio Link Control, RLC. The *Radio Link Control* (RLC) protocol provides three basic modes in Release'99 [16]:

- **Transparent mode.** The transparent mode provides basic data transfer of higher layer PDUs and optional segmentation and reassembly of upper layer PDUs *without* adding any additional overhead. Thus a transparent mode PDU is the same bit string as provided by the upper layer, no header is added. No constraints of the length regarding integer number of bytes are assumed.

  If an error occurs, other layers are responsible for error recovery and/or detection.

- **Unacknowledged mode (UMD).** Unacknowlegded mode offers the following services to the upper layer protocol:
  - Transfer of user data.
  - Optional segmentation and reassembly of upper layer PDUs.
  - Optional concatenation of upper layer PDUs.
  - Optional padding. The padding field is added at the end for reaching the predefined length of the RLC PDU.
  - Optional ciphering.

  The user data is transferred between the two endpoints in sequential order within numbered frames. No error correction is done.

  The format of the UMD PDU is plotted in Fig. 4.9(a). Because the RLC-

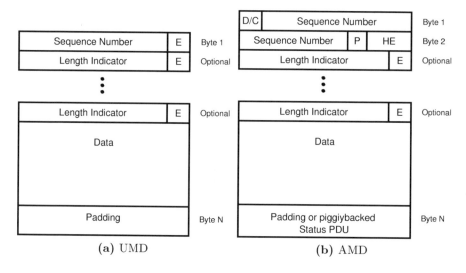

(a) UMD   (b) AMD

Fig. 4.9. Format of RLC unacknowledged mode, UMD **(a)**, and acknowledged mode, AMD **(b)**, PDUs [16]

UMD 'Sequence Number' field is 7 bits wide, all state variables in unacknowledged mode are modulo $2^7 = 128$. The UMD packets are numbered subsequently with this sequence numbers. The 'E'-Bit (Extension-Bit) indicates whether the next byte is data (E=0) or a 'LengthIndicator' field (E=1). The 'LengthIndicator' determines the length of a service data unit within the data block. Padding fills up the RLC PDU to achieve the required predefined total length.

- **Acknowledged mode (AMD).** Acknowledged mode data provides error corrected transfer of user data. An ARQ protocol detects errors and retransmits blocks in error. AMD uses negative acknowledgments to request retransmissions. A window-based flow-control mechanism (similar to GPRS RLC) is specified. The window size has a maximum range of 0 to

$2^{12} - 1 = 4095$, but it could also be lower. Acknowledged mode offers the following services to the upper-layer protocol:
- Transfer of user data.
- Optional segmentation and reassembly of upper layer PDUs.
- Optional concatenation of upper layer PDUs.
- Optional padding. The padding field is added at the end for reaching the predefined length of the RLC PDU.
- Optional ciphering.
- Error correction by using an ARQ protocol.
- In-order delivery of higher layer PDUs. If packets are received out-of-order, RLC AMD reorders the packets before handing them to the upper layer.
- Detection of duplicate PDUs, they are discarded.
- Flow control.
- Detection of errors in the protocol and recovery from such error events. The detection of erroneous blocks itself is done via the physical layer (L1) by the CRC.

The AMD PDU format is shown in Fig. 4.9(b). In addition to the already known fields from the UMD PDU, 'D/C', 'P' (Polling bit) and 'HE' fields are available. 'D/C' indicates a Data or Control PDU. 'P' is used to request status reports from the receiving RLC entity. Status and Control information PDUs in AMD mode are different from the one shown in Fig. 4.9(b). The 'HE' (Header Extension) is used exactly like the 'E' bit. The RLC-AMD 'Sequence Number' has a size of 12 bits, allowing sequence numbers from 0 to $2^{12} - 1 = 4095$. The padding field may optionally include a piggybacked status PDU to avoid explicit status messages.

Timers secure detection of error situations and the sending of regularly status reports.

In future releases hybrid ARQ schemes will also be considered in the RLC layer.

Packet Data Convergence Protocol, PDCP. The *Packet Data Convergence Protocol* is the interface from the UTRAN protocols to the network layer protocol. It is defined in [11]. Currently the IPv4 and IPv6, PPP and OSP[3] network layer protocols are supported. New network layer protocols should be possible to be introduced without changes to lower-level UTRAN protocols. The following functions are performed:

- Transfer of user data from network layer to RLC layer and vice versa. Data transfer is supported over transparent, unacknowledged and acknowledged RLC mode. In AMD RLC mode the PDCP layer receives acknowledgments for successful data transfer.

---

[3] The *Octet Stream Protocol* (OSP) is used to carry an unstructured octet (character) stream between the MS and GGSN. OSP has no existence outside the UMTS network. It has packet assembly and disassembly function and end-to-end flow control.

- Header compression and decompression of IP data streams. Standard IETF header compression schemes as defined in [43] are applied. The used header compression method is defined at the beginning of a transfer. Radio Resource Control (RRC) negotiates the algorithm and their parameters.
- PDCP sequence numbering when lossless SRNS relocation is supported. PDCP sequence numbers range from 0 to 65535. SRNS relocation is explained in Sect. 4.2. Lossless SRNS relocation is only applicable with RLC AMD mode. In the case of relocation, the PDCP SDUs are forwarded to the new SRNS. If packets arrive due to this forwarding out-of-order or multiple times the receiving entity can detect and correct this by using the PDCP sequence numbers. However, in Release'99 all maintained header compression state information is lost during relocation. Only the negotiated parameters and algorithms are saved. So the first 'compressed' packet after SRNS relocation is a full header packet. For Release'00 compression state relocation may also be moved to the new SRNS.
- Multiplexing of different radio bearers (on network level) to a single RLC entity (not yet supported but will be included in Release 2000).

PDCP PDU formats are shown in Fig. 4.10. There exist two formats.

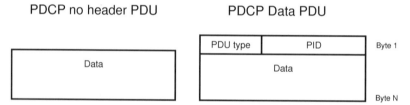

Fig. 4.10. Format of PDCP PDUs. The *left-hand side* one does not introduce any overhead and is for plain data transmission. The *right-hand side* one (Data PDU) is used with active data compression [11]

The first one does not introduce any overhead and is used for plain data transmission. No header compression is available. The second format (Data PDU) includes a PDU type field which is 3 bits in size. Currently this must have the value 000 which determines the next field (PID) to be used for header compression information. The PID field is 5 bits wide and specifies the used header compression algorithm. No data compression is specified.

Protocol termination. The termination of each protocol level in the UTRAN depends on the logical channel [15]. For a dedicated channel in the user plane the protocol endpoints are shown in Fig. 4.11. Only the physical layer spans from UE to NodeB, the Layer 2 protocols span from UE to the Radio Network Controller (RNC). In case macro-diversity combining and splitting is applied in FDD mode, part of the physical layer also spans until the RNC. If no macro-diversity is used, the physical layer is completely terminated in the NodeB.

DCH, User plane

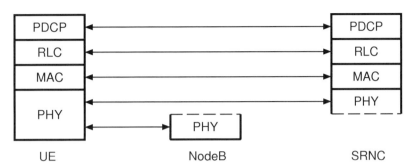

Fig. 4.11. Termination of the radio interface protocols in the UTRAN in the user plane for a dedicated channel (DCH) [15]

DTCH, User plane

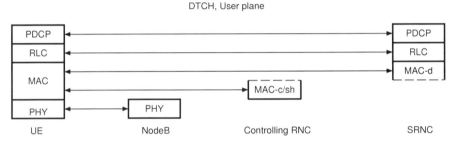

Fig. 4.12. Termination of the radio interface protocols in the UTRAN in the user plane for a dedicated traffic channel (DTCH) if controlling and serving RNC are not identical [15]

Another interesting case is illustrated in Fig. 4.12. If a UE has connection to two different Radio Network Subsystems, i.e. two different Radio Network Controllers (RNC), one RNC is the serving RNC, and the other RNC is the controlling RNC. In this case the MAC layer is operational in *both* RNCs. All common and shared channel MAC layer functions (MAC-c/sh) are terminated in the controlling RNC. The dedicated channel MAC-d functions, on the other hand, span to the serving RNC.

Protocol overhead. The total protocol stack from TCP down to physical layer is illustrated in Fig. 3.15 The data packet from the layer on top of TCP (typically the application layer) is framed and segmented into smaller blocks. Every protocol adds its own header, which increases the protocol overhead. A quantitative evaluation is plotted in Fig. 4.14. No IP protocol fragmentation is assumed here.

We plotted the overhead for five different bearer services. The overhead is largest for the 12.2 kbit/s bearer because it has the smallest transport block size (see Table 5.12 for the actual transport block size). The 2 Mbit/s bearer

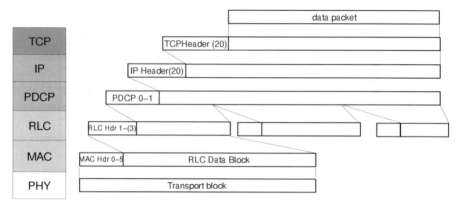

Fig. 4.13. Illustration of the packet-framing procedures done in the UMTS protocol stack

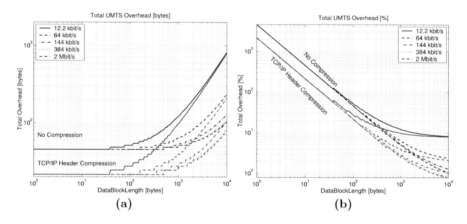

Fig. 4.14. Evaluation of the total UMTS protocol overhead per data packet for varying sizes of the input data packets (from prototol on top of TCP). Absolute values are plotted on the left-hand side plot **(a)**, relative values on the right-hand side **(b)**. See also Fig. 4.13

has the largest transport block size, thus the overhead is the lowest one. The overhead remains constant up to a data packet size of 40 to 500 bytes (for 12.2 kbit/s to 2 Mbit/s). TCP/IP header compression reduces the overhead significantly.

When we compare this evaluation with those for GPRS (Fig. 3.16), the following can be observed: Total UMTS overhead is in general smaller. The main reason for that is one additional layer (LLC) in GPRS compared to UMTS. Also the threshold data packet size, where the overhead starts to increase, is significantly higher in UMTS. This means that longer data packets introduce fewer overhead load in UMTS than in GPRS. A major cause for this effect is again the lower total overhead in UMTS.

Summarizing, UMTS overhead is almost 1/3 to 1/2 lower than GPRS overhead for equal data packet size.

4.4 UMTS Bearer Services and QoS

A bearer service provides capability for information transfer between access points [14]. Only low layer functions are involved. Packet-switched and circuit-switched domains provide a specific set of bearer services. UMTS bearer services are specified between the mobile termination MT and the core network gateway (see Fig. 4.15). The Mobile Termination (MT) func-

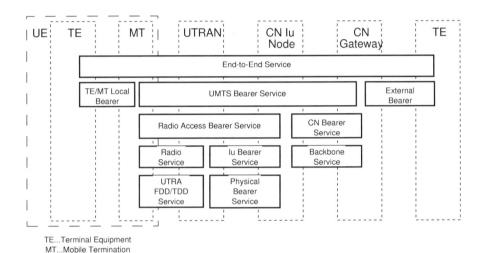

TE...Terminal Equipment
MT...Mobile Termination

Fig. 4.15. Bearer services in UMTS [14]

tion is part of the UE together with the Terminal Equipment (TE). The end-to-end service ends in the TE. The bearer services shown in Fig. 4.15 represent different levels of quality of service (QoS). The end-to-end service on the application level uses the bearer services of the underlying networks. The UMTS bearer service provides the UMTS QoS. It uses the bearer services of lower layers like the Radio Access Bearer (RAB) or the Core Network Bearer.

UMTS defines four QoS classes (or traffic classes) [14]:

- **Conversational class**. This class is characterized by a low transfer time and a maximum transfer delay. Maximum transfer delay is given by the human perception of audio and video conversation. This class includes ordinary telephony speech, Voice-over IP speech and also interactive video conferencing. The time relation between information elements must be preserved.

- **Streaming class**. The variation between information elements must be very low in this class, however, there are no requirements on low transfer delay. This class is intended for use with real-time audio and real-time video streams. But this is one-way transport only.
- **Interactive class**. The fundamental QoS characteristics are that payload content is preserved and data is transferred in a request response pattern. It is used for web browsing applications, or interactive data retrieval processes from a server.
- **Background class**. This is the lowest QoS class. The only requirement is to preserve the payload content. No time constraints are specified. This scheme applies to every kind of Internet background traffic like E-mails or downloads of databases.

The first two classes belong to the real-time traffic group (RT), the latter two classes represent non-real-time (NRT) services. A number of bearer service attributes are specified for the UMTS bearer service and the Radio Access Bearer service for all four QoS classes. The most important attributes are summarized in Table 4.3 for UMTS bearer services. The maximum bit rate for all classes is 2 Mbit/s because the radio interface is limited to this value.

The SDU error ratio is, in essence, the Frame Error Rate of SDU blocks. A number of residual BER and SDU error values are specified. Thus the QoS in each class can vary in wide ranges although basic QoS parameters are predefined due to the traffic class. The BER (and SDU error ratio) could be higher for the RT services (from 5×10^{-2} to 10^{-6}) but for NRT services lower values are specified. Transfer delay indicates the maximum delay for 95% of the distribution of the delay for all delivered SDUs during the lifetime of a bearer. It is only specified for RT classes. Also, the guaranteed bit rate is only specified for the RT classes.

The range of values for Radio Access Bearer services is the same as for the UMTS Bearer service, with the exception that the transfer delay for the conversational class is 80 ms – maximum value. The exact service attribute values depend finally on the wanted or negotiated bearer service.

Circuit-switched bearer services. In UMTS there are currently five circuit-switched bearer services (or service categories) defined [5]:

- Speech.
- Transparent data for support of multimedia.
- Transparent data.
- Non-transparent fax.
- Non-transparent data.

Each UMTS bearer is supported by a Radio Access Bearer (RAB). The RAB and the UMTS bearer are described by their QoS parameters. The bearer service parameters for non-transparent and transparent data services are listed in Table 4.4. For conversational class the maximum bit rate is always equal to

Table 4.3. Value ranges for UMTS Bearer Service Attributes [14]

Traffic class	Conversational class	Streaming class	Interactive class	Background class
Maximum bit rate [a] (kbit/s)	<2048 [b]	<2048 [b]	<2048 – overhead	<2048 – overhead
Maximum SDU size (bytes)		<=1500 or 1502 (PPP only)		
Residual BER	5×10^{-2}, 10^{-2}, 5×10^{-3}, 10^{-3}, 10^{-4}, 10^{-6}	5×10^{-2}, 10^{-2}, 5×10^{-3}, 10^{-3}, 10^{-4}, 10^{-5}, 10^{-6}	4×10^{-3}, 10^{-5}, 6×10^{-8} [c]	
SDU error ratio	10^{-2}, 7×10^{-3}, 10^{-3}, 10^{-4}, 10^{-5}	10^{-1}, 10^{-2}, 7×10^{-3}, 10^{-3}, 10^{-4}, 10^{-5}	10^{-3}, 10^{-4}, 10^{-6}	
Transfer delay (ms)	100 – max. value	250 – max. value	–	–
Guaranteed bit rate (kbit/s)	< 2048	<2048	–	–

[a] The granularity of the bit rates shall be limited to restrict the complexity of charging, interworking functions and terminals. This is for further study.
[b] A bit rate of 2048 requires transparent RLC protocol mode.
[c] Values are derived from CRC lengths of 8, 16 and 24 bits on layer 1.

Table 4.4. Circuit-switched UMTS Bearer service attributes [4,5]

Service	Transparent data including multimedia	Non-transparent data
Traffic class	Conversational	Streaming
Maximum bit rate (kbit/s)	= guaranteed bit rate	14.4, 28.8, 57.6
Guaranteed bit rate (kbit/s)	28.8, 32, 33.6, 56, 64	14.4
Maximum SDU size	280...640 bits	576 bits
Transfer delay	<200 ms	<250 ms

the guaranteed bit rate. This is the constant bit rate of the negotiated bearer service. This service is only possible in transparent mode, because no underlying ARQ protocol should influence the net bit rate. For non-transparent data service the guaranteed bit rate is always 14.4 kbit/s. The guaranteed bit rate remains on this level, even for higher maximum bit rates.

For the same reason (ARQ protocol for non-transparent service) the allowed transfer delay is higher for non-transparent service than for transparent mode.

When the wanted air interface user rate is below 14.4 kbit/s, the maximum bit rate is nevertheless set to 14.4 kbit/s.

For non-transparent services the enhanced RLP (Radio Link Protocol, as described in Sect. 3.1.2) is used for reliable communication. An SDU size corresponds to one RLP frame (576 bits).

Packet-switched bearer services. UMTS packet-switched bearer services have the same reliability, delay and throughput classes as already defined in GSM Phase 2+. Thus the Tables 3.1, 3.3, 3.4, and Table 3.5 are valid also for UMTS packet-switched data.

4.5 Differences from GSM-GPRS

GSM GPRS and the UMTS packet-switched domain are very similar. The Release'99 core network architecture is almost identical to the GSM GPRS core network architecture. The access network, however, is different. Thus a new interface has been defined (*IuPS*) and some changes to the locations of various functions have been made. In particular, the following main differences of protocol functions, besides the completely new access network, can be found:

- In GPRS a logical link for a connection is maintained by the Logical Link Control protocol (LLC). LLC spans from the mobile station to SGSN. Also the Subnetwork Dependent Convergence Protocol (SNDCP) spans from MS to SGSN (Fig. 3.11).

  In UMTS the logical link ends already in the UTRAN, specifically in the Radio Network Controller (RNC). The Packet Data Convergence Protocol (PDCP) ends already at RNC. Between RNC and 3G-SGSN the GPRS Tunneling Protocol (GTP-U) is used. No logical link is established between a 3G-SGSN and a mobile station (User Equipment, UE) any longer (Fig. 4.6 and Fig. 4.11).

  Thus a 3G-SGSN no longer has logical link management functions.
- In GSM GPRS the SNDCP protocol also defines compression of user data and not only header compression. Currently, ITU-V42bis compression algorithm is supported.

  In UMTS *no* compression of user data is supported by PDCP. Only header compression is defined. This decision is motivated by the fact that data compression efficiency depends on the type of user data. Also many applications compress the data before transmission. As compression does not work well for already compressed data it would be necessary to check the data stream whether it is already compressed or not. Also the type of possibly uncompressed data should be known for choosing an optimal compression algorithm.

  Thus, and also because compressing all user data would require a large amount of processing power, this feature has been dropped in UMTS.

- Header compression is done in the SGSN in GSM GPRS by the SNDCP protocol.
  Because the corresponding PDCP protocol in UMTS already ends in the UTRAN (RNC), the RNC has to provide processing power for all header compression algorithms used in PDCP.
- Ciphering of user data to provide confidentiality is also moved from SGSN to the RNC entity. Previously LLC provided this feature (in the mobile station and the SGSN). In UMTS, ciphering is done by the MAC layer or the RLC layer in the RNC element. For transparent RLC layer operation ciphering is done in the MAC layer, otherwise ciphering is realized in RLC layer.
- Radio Resource Management functions have been moved from SGSN to UTRAN.
- RLC-AMD and RLC-UMD have different RLC protocol overhead in UMTS. RLC acknowledged mode and RLC unacknowledged mode in GSM-GPRS have *equal* overhead.

In general, UMTS has more functionality within the access network than in the core network. Many functions have moved from SGSN to RNC compared to GSM GPRS. Thus UTRAN is more decoupled from the core network than was the case in GSM GPRS.

5. TCP in Wireless Environments

The Transmission Control Protocol is *the* transport layer protocol for reliable data transmission in the Internet. The original specification [111] dates back to 1981. Although many enhancements have been made over the years, the basic protocol assumptions are still valid. Almost every reliable network service uses TCP as underlying transport protocol. As wireless Internet access becomes more and more popular and feasible, it is a straightforward approach to use TCP also over wireless links. Furthermore, the widespread usage of TCP in the ordinary networking and Internet access almost demands usage of TCP to remain compatible with existing systems.

But TCP was designed for conventional, highly reliable links. In contrasr, wireless links are everything but highly reliable. Thus, many problems arise for TCP over wireless links.

In the following section we will address the problems arising from the usage of TCP in wireless environments. Then we will present an overview of suggested solutions already available. Performance of TCP over wireless links depends on many TCP parameters, one of them is the TCP segment length. We will derive the optimum TCP segment length for maximum throughput performance in Sect. 5.3. Sect. 5.4 includes results from simulations and measurements of data transmission with TCP/IP over GSM. This is followed by results for the packet-switched extension to GSM, GPRS (Sect. 5.5). Finally, performance of TCP/IP over the future 3rd generation mobile radio system UMTS will be presented (Sect. 5.6).

5.1 General Problems

We described the TCP protocol already in Sect. 2.3.1. In the following section we will emphasize the TCP behavior when operating on wireless links.

5.1.1 TCP Behavior on Packet Loss

As already briefly mentioned, TCP was designed to operate over highly reliable links and stationary hosts. It assumes all packet losses to be caused by network *congestion*.

TCP performs well over such networks by adapting to end-to-end delays and to network congestion. It maintains a *congestion window* which is adapted according to the current network conditions. The congestion window determines the number of allowed outstanding (i.e. unacknowledged) packets. It characterizes the amount of injected traffic into the network. A large congestion window allows intermediate buffers to be filled during a connection regardless of the actual available bandwidth. The sender simply transmits all packets within the congestion window without waiting for any acknowledgment. This in turn makes it possible to react rapidly to changes in network behavior, because the packets in the buffers can be sent immediately if bandwidth increases.

Fig. 5.1. Sample network with TCP data transmission from Host A to Host B via a bottleneck link

Consider a simple example as drawn in Fig. 5.1. TCP is used for data transmission from Host A to Host B via a bottleneck link between Node C and Node D. The bandwidth of the bottleneck link is small compared to the connections between Host A and Node C and also between Node D and Host B. The *Round Trip Time* (RTT) defined as the time a packet needs to reach the endpoint and travel back again, is assumed to be rather high (>0.5 s). We assume a filled buffer before Node C. If the bottleneck bandwidth becomes larger (e.g. due to additional available network capacity), packets stored in the buffer can be immediately sent over the bottleneck link towards Host B. Then Host B is able to send acknowledgment packets faster to inform the sender about the higher available bandwidth (via the higher acknowledgment rate).

If the buffer is not filled, and the bandwidth becomes larger, packets already in the network between Host A and Node C are the *only* available packets for sending over the bottleneck links. However, these packets arrive at the rate of the previous bottleneck bandwidth (because the rate is still given by the rate of the acknowledgments coming from Host B). Thus the TCP sender needs at least one RTT to get informed about the higher possible bandwidth. Thus a filled buffer gives better network utilization. In fact, the TCP Sender always probes for higher possible bandwidth in the 'congestion avoidance' state, only the time it needs for detection differs if the buffer is filled or not filled.

To fill the buffer, a large congestion window is necessary. The larger the RTT, the larger the congestion window should be. A small congestion window implies slower reactions to changes and more dependence on the RTT. A good estimate for the congestion window in stationary state is cwnd = BW × RTT, where BW is the bottleneck bandwidth.

When filled, large buffers are important for good network utilization. However, the packet delay is also increased with a larger buffer length. If the network is in stationary state, a large buffer forces the packet to remain a longer time in the buffer. This might not be acceptable for real-time or interactive traffic. Thus a compromise between good network utilization and packet delay must be found.

If a packet gets lost on its way from Host A to Host B, TCP assumes a congestion in between A and B. This could be caused by buffer overflow in the buffer at Node C due to a temporarily smaller bottleneck bandwidth. The packets already on the way from A to C arrive at the fast rate as had been appropriate before the bandwidth reduction. But the packet rate out of C is reduced. Thus the buffer might not absorb all packets, and some might get lost. When TCP notices this packet loss, it reduces the congestion window to reduce the output rate of new packets.

The exact behavior is slightly different in different TCP versions, but the effect stays the same, more or less pronounced. In Fig. 5.2 the congestion

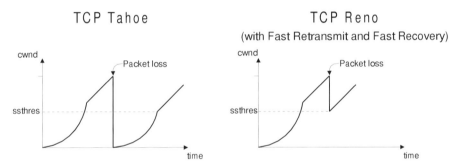

Fig. 5.2. Adaptation of the congestion window, cwnd, at the startup phase of a connection and in the case of a packet loss. The window adaptation is plotted for both TCP versions, Tahoe and Reno

window, cwnd, adaptation for TCP Tahoe (without Fast Recovery) and for TCP Reno (with Fast Recovery) is sketched. If a packet loss occurs, TCP Tahoe sets back the congestion window to the initial value as it would do at the beginning of a new connection. Afterwards, the 'slow start' algorithm is executed which exponentially increases the congestion window over time. TCP Reno acts similarly, but it sets cwnd down only to a threshold value, ssthres. ssthres is set to one half of the congestion window before the packet loss. Also no 'slow start' is done but the 'congestion avoidance' algorithm is executed, which increases cwnd linearly with time.

Both versions decrease the injected amount of traffic into the network dramatically. When the packet loss has been caused by congestion, this reaction is appropriate. However, if the loss has been only a single isolated one (e.g. caused by a link error), this answer is *not* appropriate. The traffic injection is dropped not only for a short amount of time but for a considerable number of packets after the packet loss. In the ideal case, the congestion window should be dropped *only* if a network congestion occurs. If the packet loss has been caused by a transmission error, cwnd should remain at the current level. To maintain the number of packets in transit constant the pause should be just the time needed for a single packet (for the last one). This ideal behavior is illustrated in Fig. 5.3. At time t_1 and t_3 a packet loss occurs due to

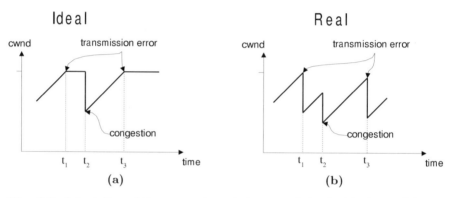

Fig. 5.3. Adaptation of the congestion window, cwnd, in the ideal case **(a)** and the real case **(b)**, TCP Reno

transmission errors. At time t_2 a packet loss caused by congestion (e.g. buffer overflow) is assumed. In the ideal case (Fig. 5.3(a)), cwnd is only dropped for the congestion case. In reality, however, the congestion window is dropped for every packet loss, regardless of the cause of the error (Fig. 5.3(b)).

This behavior is very poor in *mobile, wireless environments*. In such systems, *most* of the packet losses are caused by transmission errors. This severely violates the assumptions TCP was designed for. Every error detected by TCP causes a congestion window drop. This in turn reduces the amount of data injected into the link, which decreases the actual throughput unnecessarily. The link quality in mobile, wireless data transmission systems is usually rather poor, especially for transparent data transmission. In transparent mode, TCP is the lowest protocol layer providing reliable transport service. Thus TCP catches *all* errors. Bandwidth is wasted simply by wrong reaction of TCP to transmission errors and the wireless link can not be utilized up to its nominal bandwidth.

5.1.2 TCP Behavior on Temporary Disconnections

In mobile, wireless environments, temporary disconnections can occur frequently. Possible causes are handover failures between cells, shadowing of the radio path or simply a temporary poor link quality leading to a high BER caused by short-term fading, high mobile station velocity, high co-channel interference or high path loss at cell borders.

Disconnections caused by functions executed by the mobile radio system itself have different effects on the data stream than that caused by link errors. In any case, we have to distinguish between system-managed disconnections and disconnections due to poor link quality. In the following we will discuss both cases.

One effect of frequent disconnections, either handovers or disconnections due to link errors, is the exponential increase of the TCP retransmission timer. We will discuss this effect at the end of this section.

- **System managed disconnections, Handover.** A variety of handovers can occur. Handover between two cells or two different frequencies in GSM is the classical example. Two cases can be distinguished:
  - **Non-lossy handover**. The handover causes only a temporary disconnection visible to the data stream. However, *no* data is lost due to the handover. All packets sent by the sender are delivered to the receiver (assuming no other transmission errors). Thus the only effect TCP perceives is an increase in delay. No retransmissions must be scheduled, the congestion window is *not* reduced, and the data stream is only delayed. But, if the delay caused by the handover exceeds TCP's retransmission timer, TCP schedules a retransmission nontheless. TCP reacts as if an error would have occurred, although this is not the case.
  - **Lossy handover**. In the case of lossy handovers, the mobile network may not deliver all data. Some packets can get lost simply because of the handover. For example, the mobile station switches to another base station and the data already on the way to the previous base station will not be resent to the new base station. In that case, TCP perceives not only a delay but also a packet loss. This leads to retransmissions and additional traffic.
- **Disconnections due to link errors.** Disconnection due to link errors are always lossy. That is, packets get lost and must be retransmitted by higher layers (link layer or TCP). The disconnection time depends on the actual cause.

The reaction of TCP to such disconnections is, again, to reduce the congestion window (if packets get lost or the retransmission timer expires). In the ideal case, the TCP sender would know that a disconnection had been the cause for packet loss or delay and only interrupts sending of packets for the time of disconnection. The ideal and real behavior for disconnections is sketched in Fig. 5.4. The effects in the real case have already been pointed out in the previous section for packet loss.

Exponential increase of retransmission timer. The TCP retransmission time is doubled every time a retransmission is scheduled, i.e. by expiration of the TCP retransmission timer. This is the back-off algorithm,

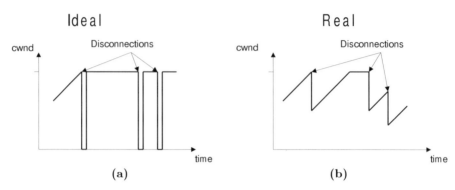

Fig. 5.4. Adaptation of the congestion window, cwnd, in the ideal case **(a)** and the real case **(b**, TCP Reno) for temporary disconnections

known to remain stable. If packet loss occurs purely accidentally during the transmission of a specific packet and also on its subsequent retransmission attempts, the retransmission timer becomes relatively large. See Fig. 5.5 for an example. The TCP sender receives an acknowledgment (ACK#0) and sends thereafter the Packet #1 (Xmit#1). Packet #1 never reaches the receiver due to a temporary disconnetion, thus the retransmission timer expires at the sender. The packet is retransmitted (first Rexmit#1). This retransmission gets lost again. But the sender has already doubled the retransmission timer, a procedure which increases the timer exponentially.

This exponential increase of the retransmission timer is the actual source for long idle times in data transmission with frequent errors or disconnections.

5.1.3 Interactions with Link Layer

In non-transparent data transmission modes an additional link layer protocol is used to correct errors on the radio interface. In GSM-CSD the *Radio Link*

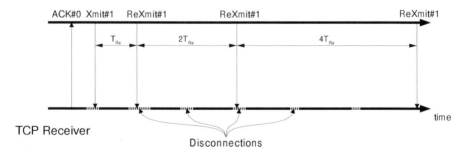

Fig. 5.5. Exponential increase of the TCP retransmission timer in the case of frequent retransmission timeouts due to temporary disconnections

Protocol (RLP) performs this task. GSM-GPRS and UMTS have similar protocols in the *Radio Link Control* (RLC) layer. All these protocols use the *selective repeat* ARQ technique. Retransmissions of packets in error are used for error correction. Thus the throughput, and especially the packet delay, depends on link quality.

In these non-transparent modes two ARQ protocols work independently of each other on different protocol layers, TCP and RLP (or RLC). As TCP relies on lower level service, thus also on the link layer service, several interactions between TCP and the radio link layer protocol can be identified:

RTT estimation. In TCP, the estimation of the Round Trip Time (RTT) is done dynamically for every packet, that is acknowledged *without* retransmissions. A running average of RTT is maintained in the TCP sender.

As outlined above, the packet delay varies with the link quality in non-transparent data transfer mode. Due to link-quality-dependent measurement snapshots also the RTT estimation becomes unreliable.

Subsequently, the retransmission timeout value, T_{Re}, is calculated from the estimated mean round trip time,

$$T_{\mathrm{Re}} = K\sigma_{\mathrm{RTT}} + \overline{\mathrm{RTT}} \, , K = 2 \ldots 4 \, , \tag{5.1}$$

where σ_{RTT} is the standard deviation of the round trip time and $\overline{\mathrm{RTT}}$ its mean value. Thus also the retransmission timer is subject to the RTT estimation problem.

If the link quality covers a wide range or has frequent occurrence of bad states, σ_{RTT} will become larger. This leads to a large retransmission timer, T_{Re}. A large T_{Re} implies slower reaction to packet losses which may cause throughput degradation. The authors of [25] report that high values of T_{Re} reduce the throughput significantly because the reaction of TCP upon packet loss becomes slower.

The packet delay variation due to link layer ARQ protocols packet retransmissions has been identified in [33] also for Wireless LAN (WLAN) type radio access systems. Although the link layer protocol was able to reduce the residual packet loss rate to about 10^{-6}, wrong TCP RTT estimation led to severe TCP performance degradation.

However, in this case the RTT estimation is misled the other way round from that described above, i.e. it is underestimated. Thus in such cases T_{Re} is smaller than the actual TCP/IP packet delay which leads to a timeout. This problem is reported in [33] to be more severe for TCP implementations having a very fine timer resolution ($\approx 10\,\mathrm{ms}$). But bear in mind that this was found for a WLAN with very high bit rate ($\approx 10\,\mathrm{MBit/s}$) and not in a mobile telephone network like GSM with comparable low bit rate and high latency.

An accurate RTT estimation is also among the reasons reported in [95] that TCP performs well over GSM RLP. When the time stamp option is used, i.e. every packet can be used for RTT update, competing retransmissions between TCP and RLP are rare (see also below).

Link Layer out-of-order delivery. If the radio link layer protocol does not provide in-order delivery of packets, packets may reach the TCP receiver *out-of-order*. As an example, the same packet could have gone lost multiple times and some error counter in the link level protocol requests not to wait any longer for this packet. But the packet is finally received by the radio link receiver and eventually delivered to the upper layers, but out-of-order.

Out-of-order packets reaching the TCP receiver cause sending of duplicate acknowledgments to the TCP sender. This in turn requests the TCP sender to invoke fast retransmission and fast recovery algorithms. This action potentially degrades the throughput of the link, especially when the delay-bandwidth product is large[1]. Because at large bandwidth-delay links, many TCP packets are in transit and TCP's reaction needs some time to be effective.

Timer interaction and competing retransmissions. TCP implementations like TCP Tahoe and TCP Reno have a coarse timer granularity of 500 ms. On the other hand, radio link layer protocols usually have finer timer granularity, system dependent, but within a millisecond time scale.

If the link layer protocol is a low performance ARQ implementation (like stop-and-wait) the TCP retransmission timer may expire before the link layer protocol has recovered from a packet loss [45]. So essentially both protocols, TCP and the radio link layer protocol retransmit the lost data. These are called competing retransmissions.

For high-performance radio link protocols with selective repeat ARQ, this problem is reported to be not as serious [149].

Similar results can be found in [95]. There it is found by measurements that TCP and GSM RLP rarely interact in an inefficient way. But this seems not to be true for every TCP implementation. The TCP implementation used in [95] (BSDi 3.0) shows almost no interactions while others, having a more aggressive retransmission timer ($K = 2$ in (5.1)) do have interactions.

5.1.4 Asymmetric Services

Internet traffic such as Web browsing causes heavily asymmetric traffic. The downlink traffic is typically orders of magnitudes higher than the uplink traffic. GPRS and UMTS take this into account by being prepared for asymmetric configuration.

In TCP, the acknowledgment rate arriving from the TCP receiver drives the sending rate. This is the 'self-clocking' property of TCP (see Sect. 2.3.1). But acknowledgments also require bandwidth. If we consider only downlink data transfer, the acknowledgments are the only 'data' sent in the uplink. But if the uplink bandwidth is too small for the required acknowledgment rate, the sending rate of new data packets in the downlink may be limited due to the limited bandwidth in the *uplink*.

---

[1] A large delay-bandwidth product represent a large number of packets in transit.

We will now give a simple example considering TCP/IP packet overhead (20 + 20 bytes) without header compression and without taking lower levels into account (Table 5.1). We will assume a GPRS system with 8 timeslots allocated in the downlink for the actual data transfer, and 1 timeslot in the uplink for acknowledgments. Also we will investigate a UMTS TDD system in principle. UMTS TDD provides 15 timeslots per radio frame. At maximum 14 timeslots can be allocated for one link direction (in this example for the downlink) and the remaining one for the opposite direction (here the uplink). Several different timeslot formats are defined [12]. If we assume downlink timeslot format #19 from [12], a code rate of 1/3 (turbo coding) and a total protocol overhead of 10% to 7% (400 bytes to 1000 bytes packets, see also Fig. 4.14) the user data rate per time slot is about 131.5 kbit/s to 135.9 kbit/s. In the uplink, we assume slot format #97, code rate 1/3 and 10% protocol overhead (only small acknowledgment packets). Hence, an uplink user data rate per timeslot of about 132 kbit/s is considered. In this consideration of UMTS-TDD *no* additional overhead for logical channels or system management is included, i.e. the data rate estimation is only approximate.

The total data rates are specified by the allocated time slots for uplink and downlink in both systems. Assuming TCP payload length of 400 and 1000 bytes, the downlink TCP/IP packet rate can be computed from total packet length and downlink bandwidth (DL_PacketRate = DL_Bandwidth/BitsPerDataPacket). If every TCP packet causes a TCP acknowledgment, the required acknowledgment rate in the uplink is equal to the maximum packet rate in the downlink. Thus the required uplink bandwidth is given by UL_Bandwidth = DL_PacketRate · BitsPerAckPacket.

The results for all parameter combinations are summarized in Table 5.1. For the GPRS system and low TCP packet length (400 bytes) almost all available uplink resources are needed for acknowledgment packets for the downlink data transfer. Thus only a small amount remains for actual uplink user data transfer.

In the UMTS-TDD example, the 1000 byte packet transmission requires 57% of the available uplink resources. But the downlink data transmission of 400 byte packets would require significantly *more* bandwidth in the uplink than actually available (184 kbit/s compared to 132.1 kbit/s).

An improvement of this situation can be achieved by several mechanism:

- Acknowledging only every r^{th} received TCP data packet, known as 'delayed ACK' mechanism. This reduces the required uplink bandwidth by the factor r. Typically $r = 2$, so the reverse bandwidth is halved. A problem is the reception of less then r packets, because then *no* acknowledgment is sent at all. A timeout has to clear this situation, leading to degraded performance at slow start.

Table 5.1. Examples of required uplink bandwidth just for acknowledgment traffic for TCP downlink data transfer for various system parameters of a GPRS and a UMTS-TDD system

| System arrangement | GPRS (8 TS DL / 1 TS UL) | | | | UMTS TDD (14 TS DL / 1 TS UL) SF=1 | |
	CS-1		CS-4			
TCP Payload length (bytes)	400	1000	400	1000	400	1000
Downlink bandwidth (kbit/s)	9.05×8 $= 72.4$		21.4×8 $= 172.2$		131.52×14 $= 1841$	135.9×14 $= 1902$
Uplink bandwidth (kbit/s)	**9.05**		**21.4**			**132.1**
Downlink packet rate (Packets/s)	20.6	8.7	48.9	20.7	575	237
Uplink (Ack) packet rate (Packets/s)	20.6	8.7	48.9	20.7	575	237
Required UL bandwidth (kbit/s)	**6.58**	2.78	**15.64**	6.62	**184**	75.8

- The *Ack Congestion Control* (ACC) algorithm [27] employs a dynamically varying delayed-ack factor r. It is based on the congestion of acknowledgments in the reverse link which is detected by the *Random Early Detection* algorithm at the gateway [72]. The problem here is the detection of transmission errors and the behavior of r when transmission errors occur. r is only changed on congestion, which is not detected when transmission errors occur.
- The *acknowledgment filtering* method developed in [27] is based on controlling the amount of acknowledgment packets at the reverse link router. The router scans the acknowledgment packets. If a new acknowledgment is received, a fraction of the acknowledgments belonging to the same connection is discarded. Because TCP acknowledgments are cummulative only the last acknowledgment is necessary to transmit. Thus the acknowledgment rate is reduced.

Nevertheless, one has to bear in mind that acknowledgment traffic still needs considerable resources. In some cases (see UMTS-TDD example above), acknowledgment traffic reduction is even mandatory to utilize the available bandwidth efficiently.

5.2 Suggested Solutions from Literature

Many solutions have been proposed over the last few years for improving the performance of TCP when operating over wireless links. According to [28] we can classify the schemes into three basic groups:

- End-to-end
- Split-connection
- Link-layer

We will present the concepts of these three techniques and cite examples for each group.

5.2.1 End-to-end proposals

End-to-end proposals try to make the TCP sender and receiver endpoints aware of wireless problems. The improvements are restricted to the endpoints. The transmission line and the lower-level protocols are not affected at all. One can think of many enhancements, some are only minor and others are indeed major changes in the TCP protocol itself.

Fast retransmission and fast recovery. A very effective method is already included in new TCP versions (TCP Reno and TCP Vegas). It is the 'Fast Retransmission' together with the 'Fast Recovery' algorithm. In [35] Fast Retransmission is considered as an effective solution to resume communication immediately after a handover without waiting for a retransmission timeout.

In [35] an additional step is also considered. TCP is informed by the lower layers about handovers. Thus TCP can immediately provide a fast retransmission if the handover has been completed.

However, this extends the method beyond end-to-end, because handover issues are only known directly at the wireless link endpoints. Usually, neither TCP sender nor receiver is likely to be located at a location, where handover information is available. Thus lower layers and intermediate entities would also have to be modified.

Selective Acknowledgment TCP option. In [97] a selective acknowledgment enhancement for TCP is specified (SACK). It is implemented with additional TCP options. The selective acknowledgment mechanism is combined with a selective repeat retransmission policy.

In the case of multiple losses within a transmission window of data, TCP may experience poor performance. Because conventional TCP uses cumulative acknowledgments, the sender unnecessarily retransmits segments which have been received correctly. After the TCP endpoints have signalized that they support TCP SACK, the SACK option is sent by the receiver to the sender. As for every TCP option, it is located in the *Options field* of the TCP header. It contains the edge-sequence numbers of non-contiguous blocks already successful received. These non-contiguous blocks of segments are interrupted by not yet received segments that need to be retransmitted. After receiving the SACK option, the TCP sender has to retransmit exactly the gaps between the blocks of segments already received.

If the return path carrying the SACK options is lossless, indicating one last data packet per SACK option would be sufficient. Because that is not the case in reality, the SACK option is designed to carry multiple blocks of segments already received. A SACK option specifying n blocks will have a length of $8n + 2$ bytes. Thus at maximum 4 blocks can be specified, because of the limit of 40 bytes for TCP options. If SACK is used in conjunction with other TCP options (e.g. timestamp option), the number of possible blocks is further reduced.

However, SACK introduces some pitfalls. There is a trade-off between reducing the overhead with SACK because the number of retransmitted packets is reduced, and increasing the overhead due to the increased header length. The introduced overhead is significant, the mandatory TCP header already covers 20 bytes. Every additional TCP option used leads to worse overhead performance. This depends on the error characteristic of the transmission path, if SACK increases performance or not. If the number of losses within a transmission window easily exceeds unity, SACK will increase performance despite the introduced overhead. This is the case with either high error rate and/or a large receiver window. One also has to bear in mind that the overhead is *only* introduced in the return path (at the acknowledgment packets). Thus, the situation with asymmetric traffic is made worse (see Sect. 5.1).

Another problem arises if SACK is used in conjunction with TCP/IP header compression [81]. The header compression leaves a TCP/IP packet unchanged, if TCP (or also IP) options have been changed against the previous one. Thus every time TCP options changes (or are used in a packet but not in the previous one), the packet is delivered uncompressed. Thus, if TCP-SACK options are used extensively, this could lead to serious performance degradation of the compression algorithm.

Mobility aware TCP. A few proposals introduce mobility features into TCP. This makes TCP aware of the mobility of the user. These mobility features are also considered to improve performance in non-wireless but mobile environments. The following two proposals belong to this class of TCP enhancements.

In [122] a 'Mobile TCP' protocol is proposed. Two mechanisms are introduced:

- Performance Management:
  This should improve performance of TCP in the case of handovers. The mobile host has to discover a handover and must also notify the corresponding host, so that both TCP endpoints are aware of the handover. Now the sender retransmits these packets currently in the pipe but without triggering congestion-control mechanisms. As a starting point for the newly established connection after the handover, the old values from the previous connection are used for all dynamic TCP parameters, like estimated RTT, retransmission timeout and congestion window.
- Connection Management:
  'Mobile TCP' includes mobility management function like in MobileIP but on the transport layer. This avoids the 'triangle routing' problem (see also

Sect. 2.2.3). A list of known addresses is stored at each TCP endpoint and exchanged between the endpoints by the use of new TCP options.

The performance of this approach seems to be especially advantageous if the network delay is large ($\tau > 100\,\text{ms}$). Then, more packets are already in the pipe if a handover occurs, and more packets profit from avoiding triggering the congestion control.

Another proposal, 'M-TCP' [34], deals especially with long and frequent disconnections as well as high BER. It is in fact a kind of hybrid solution: Although it is a *split connection* approach (see also below), end-to-end TCP semantic is maintained. M-TCP is only used between a serving host (i.e. the base station) and the mobile host. For the connection between serving host and the fixed host (e.g. in the Internet), conventional TCP is applied.

Because the bandwidth within the wireless link (between serving host and mobile host) is usually fixed[2], there is no sophisticated congestion control algorithm included in M-TCP. M-TCP gets notifications from lower levels about wireless link connection status and can react. When connection is lost, all timers are frozen. But the congestion window at the intermediate host is not shrunk. Also, if a retransmit timeout occurs, the segments are not retransmitted immediately but M-TCP waits for acknowledgments packets to arrive. Because it is assumed that the link is disconnected when no acknowledgment packets arrive at the serving host, it makes no sense that packets are retransmitted (even if they are timed out). This is called the 'persist mode'.

This behavior leads to good performance in the case of disconnections, especially for frequent disconnections (e.g. handover). As the sending window at the serving host is not shrinking, high BER situations can also be handled. When the link is recovering from poor BER values, the sending rate is the same as before.

End-to-end semantic is preserved, in contrast to the solutions presented in the next subsection. This is because the TCP segments coming from the fixed host and arriving at the *serving* host are *not* acknowledged until they are really acknowledged from the *mobile* host (via M-TCP).

One problem is the split connection approach (although end-to-end semantic is maintained, 'hybrid solution'), as a special serving host needs to be implemented in the base station, which violates transparency and the layer structuring in the communication system (see also Sect. 5.2.2).

A third solution, the 'Window Prediction Mechanism' introduced in [23] proposes an end-to-end mechanism based on a varying, delayed acknowledgment factor r that adapts to the amount of received packets. The receiver tries to predict the congestion window at the TCP sender. Thus the future packet rate that will be received can be estimated.

This method overcomes the shortcomings of the various delayed-acknowledgment mechanism presented in the previous section. They have poor per-

---

[2] This might not necessarily be true in packet-switched mobile radio systems like GPRS and UMTS.

formance in the case of frequent transmission errors. The window prediction method takes special considerations on duplicate acknowledgments and retransmitted packets, which occur frequently in wireless environments.

Distinguishing between congestion and loss of data. Some of the above proposals rely on the assumption that a packet loss due to errors can be somehow distinguished from losses due to congestion. If the cause for the packet loss is known, the action to be executed is almost always the same:

> 'Trigger the TCP congestion algorithms (slow start, fast retransmit, fast recovery) *only* if there is a congestion. If a loss is due to a transmission error, just retransmit the packet but do *not* decrease packet output rate at the TCP sender.'

However, this task is not easy to accomplish. For detection of transmission errors as opposed to congestion errors, the following methods can be used:

- Explicit Loss Notification [35]:
  The Sender is informed by the mobile host with a special TCP option that a previous packet has been discarded due to transmission errors. An intermediate host could also add a TCP option to the acknowledgment packets indicating that packets have been discarded due to congestion. Either the packets lost due to congestion or transmission errors are reported to the sender. Therefore the sender can distinguish between the type of packet losses and react appropriately.
- Using inter-arrival time:
  In [31] a scheme, using inter-arrival times of packets at the receiver, is proposed. It is assumed that the wireless link is the last hop to the receiver and has the smallest bandwidth among all links involved in the connection. The basic idea is the following: If a packet is discarded from an intermediate host due to congestion, there are still enough packets in buffers to hold the transfer speed. Thus the receiver may experience the same packet rate and the gap between all packets corresponds to the available bandwidth.
  However, when the packet gets lost in the wireless link, a longer gap is experienced at the receiver. The time needed for a packet transmission will expire until the next packet is received. Thus the receiver knows, because of the gap, the packet encountered a transmission error.
  The mechanism works well for bulk data transfer. But if the packets are sent not equally spaced (e.g. interactive traffic) the receiver cannot distinguish between errors or gaps due to inactivity.

5.2.2 Split-Connection Proposals

In the split-connection approach the TCP connection from the fixed host is terminated at an intermediate host (e.g. the base station). A separate TCP connection is used from the intermediate host to the mobile host. As the second TCP connection operates only over the wireless link, it can be highly optimized for this environment. Even a totally different transport layer protocol would be possible over the wireless link, although there could arise problems with standard applications at the mobile host. The optimized protocol should have improved performance over standard TCP over the wireless link. The split-connection approach is illustrated in Fig. 5.6.

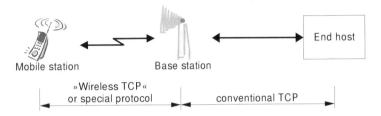

Fig. 5.6. Splitting the TCP connection into two separate connections. An optimized protocol derivate can be used over the wireless link

Two fundamental problems arise with this approach:

- The end-to-end semantic is violated. Consider a data transfer from a fixed host to the mobile host. As the first connection ends at the intermediate host, acknowledgments are generated by the intermediate host as soon as data packets arrive. Although the intermediate host will send the data packets as soon as possible to the wireless host, the original TCP sender at the fixed host may receive acknowledgments even *before* data packets arrive at the mobile host.
  Although the packets are possibly sent properly to the mobile host, there exist *no* end-to-end connection at the transport layer any longer. This contradicts the OSI-Layer model.
- Buffer space at the intermediate host may overflow. Because the TCP packets are acknowledged immediately from the intermediate host, the TCP sender at the fixed host assumes a well-working link. Thus it sends data packets at the same rate, or even higher, continuously. If the intermediate host is not able to maintain this high rate (e.g. due to errors on the wireless link or simply because bandwidth is lower), TCP packets from the fixed host have to be buffered at the intermediate host.
  Thus a well-dimensioned buffer is important for this kind of technique.

The split-connection protocols can be classified into those that violate the end-to-end semantic, and others that maintain it.

One of the first split-connection approaches was Indirect-TCP (I-TCP) [26]. I-TCP maintains two completely independent connections between fixed host and intermediate host and between intermediate and mobile host. Thus the end-to-end semantic is lost. The connection over the wireless link deals mainly with handoffs (disconnections) and is aware of such interruptions.

There are no specific modifications against high BER. Only the fact that retransmissions are only made over the wireless link and not through the whole end-to-end connection improves the performance.

In [28] the 'Snoop' protocol is presented. It consists mainly of modifications of the network layer software at the base station. The base station caches the data coming from the fixed host. It determines whether the packets arrive in sequence or not, or whether duplicate packets arrive, thus detecting congestion losses in the link from fixed to intermediate host. The packets at the intermediate host are forwarded to the mobile host. If acknowledgments from the mobile host arrive at the intermediate host they are examined again. If it is a new acknowledgment it is forwarded to the fixed host and the corresponding data packet in the intermediate host's cache is removed. If it is a duplicate acknowledgment a lost data packet is indicated. Then immediately a local retransmission (from the intermediate to mobile host only) is performed and no acknowledgment is sent to the fixed host.

In this approach the end-to-end semantic is preserved because acknowledgments are sent to the fixed host *only* when data has been successfully transmitted over the wireless link. This works well in environments with high BER because quick local retransmissions are performed. But problems could arise for long and frequent disconnections. Then, the TCP timeout mechanism at the fixed host is activated because no acks arrive from the intermediate host at the fixed host.

In fact it is reported in [115] that Snoop TCP does *not* work well in a GSM-GPRS system. Throughput is lower and the number of retransmissions are higher than with conventional TCP-Reno. The main reason is the high delay introduced at the GPRS radio interface. This causes an unfair RTO estimation at the Snoop agent which leads to duplicate retransmissions of the TCP sender at the fixed host and of the Snoop agent.

A more sophisticated elaboration of the split-connection idea has been implemented in the Mowgli[3] system [86]. The Mowgli approach involves all levels of communication. A special transport service, the *Mowgli Data Channel Service*, MDCS, transparently replaces the standard TCP/IP protocols on wireless links. A special 'Mobile-Connection Host' translates the MDCP protocol into standard TCP/IP. All connections must be routed over this host. Although performance improvement was good, changes to existing systems and applications are rather high, which minimizes the chance of an implementation in a real system.

In [121] an approach with two intermediate hosts is introduced. The connection is split into three different parts with the one going over the wireless link in the middle. On the wireless link a special *Wireless Link Protocol*, WLP, is used. It involves selective repeat ARQ and also fragmentation of large TCP/IP packets. The three connections are totally independent. End-to-end semantic is broken. This technique is not really acceptable for cellular

---

[3] Mobile Office Workstations using GSM Links.

systems. It is more efficient for satellite or directional radio systems, because two intermediate hosts are necessary.

In [114] a transmission control scheme, WTCP, is introduced which is also a split-connection method. It maintains end-to-end semantics and requires no modification of the TCP protocol at the fixed and mobile hosts. WTCP is the counterpart to the TCP protocol but only operated at the intermediate host to the mobile host. Incoming packets from the fixed host are stored and forwarded to the mobile host. Local retransmissions are performed over the wireless link by WTCP. It tries to hide the time spent for local retransmissions from the fixed host, thus the TCP RTT estimation at the source is not disturbed.

A theoretical wireless Internet access architecture is presented in [148], called AMICA: Adaptive Mobile Integrated Communication Architecture. This includes a 'Remote Socket Architecture' to provide the TCP services in the MS although TCP is terminated at the access point (base station) by a special TCP engine (split connection approach). A wireless technology independent *Export Protocol, EXP,* is introduced which maps the logical TCP service access points to the MS and holds the connection to the TCP engine in the base station. EXP spans from the base station to the MS. A lower layer *Last Hop Protocol, LHP,* covers the link layer functionality and is wireless technology dependent.

AMICA covers an abstract system view which provides TCP/IP service on any type of wireless system, but is especially developed for WLAN access networks. It is not clear from [148] if TCP end-to-end semantic is violated or not. If it is not, EXP and the TCP engine need to have mechanisms to ensure that.

However, AMICA is only a vision of a modern communication architecture including wireless access not a real system.

5.2.3 Link Layer Proposals

The Link Layer proposals try to hide the link-related losses from TCP by using local retransmission and perhaps forward error correction. The local link protocol can be tuned for operating over wireless environments. A number of link layer protocols are available. Almost every system provides a mode using such a protocol. For GSM these are the Radio Link Protocol, RLP, for circuit-switched data (Sect. 3.1.2) and the RLC layer in GPRS (Sect. 3.2.4) for packet-switched data. In UMTS packet-switched data it is again the RLC layer (Sect. 4.3.2) that provides reliable link data transfer. All of them use selective retransmissions and basic flow control schemes.

But as far as it concerns GSM RLP, its implementation might not be optimal. According to the results in [93] a TCP throughput improvement of up to 25% can be achieved by increasing the RLP frame size up to 210 bytes. In the GSM standard a fixed frame size of 30 bytes is applied. The reason is the powerful FEC in GSM CSD which makes ARQ error correction in a wide range of different environments obsolete. In [93] even an adaptive RLP frame size is suggested (30 to 210 bytes) to achieve optimum throughput (see also Sect. 5.3 and the measurement results in Sect. 5.4). This seems possible because the BLER varies slowly enough to make an adaptive error control scheme applicable. But this would include a change in the standard.

The advantage of the link layer approach is clearly that the transport layer (TCP) needs not to be changed. This is important because TCP is widely used and it is impossible to change all TCP implementations currently in use. Also all TCP enhancements based on options are not mandatory. They may or may not be implemented, especially on the fixed net host. Thus in the ideal case, TCP does not recognize that it operates over a highly unreliable link. Hence, no unsuitable reactions are done by TCP and it performs well also over wireless links.

Additionally, the strict OSI layering is preserved and no protocol snooping has to be done at higher layers.

Some problems of this approach have been already pointed out in Sect. 5.1.3. One major problem was the extensive delay variation on the IP level introduced due to link layer retransmissions. An approach presented in [70, 71] is a possible solution to that problem:

Simultaneous MAC Packet Transmission (SMPT). Although SMPT was developed for WLAN broadband radio access we mention it in this work because it relies only on a CDMA system and no further assumptions have been made. Thus it could be used, in principle, in every CDMA system.

The terminology in [71] assumes a MAC layer which ensures reliability due to retransmissions, i.e. the MAC layer includes also an ARQ protocol (e.g. like RLP in GSM). The SMPT approach assumes a CDMA-based transmission technique where multiple channels are available. Each channel is determined by its spreading code. A MS can use multiple channels (codes) simultaneously for a single connection.

The basic idea of SMPT is to retransmit corrupted or lost link layer packets on parallel channels, while the subsequent packets are transmitted on the initial channel. Figure 5.7(a) illustrates the conventional case[4] and Fig. 5.7(b) the SMPT approach. If a packet gets corrupted (e.g. packet #2) it is immediately retransmitted on a new channel *simultaneously* to the transmission of the subsequent packet #3 on the initial channel. Thus if a TCP/IP packet has been segmented into smaller link layer packets the IP layer does *not* recognize the retransmission at the link layer. Because if both packets #2 and #3 are then (re-)transmitted successfully, the overall delay of the IP packet does not change compared to the error-free transmission. Similarly if more packets get lost, more parallel channels are used for retransmissions to maintain the IP packet delay. Only in the case when the corrupted packet is the

---

[4] Here a simple stop-and-wait ARQ protocol is assumed because the sender waits for packet #2 and #4 to be transmitted successfully until the subsequent packets are transmitted.

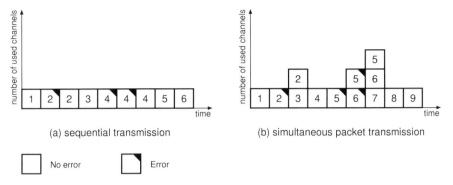

(a) sequential transmission   (b) simultaneous packet transmission

▢ No error   ◤ Error

Fig. 5.7. Conventional link layer packet transmission (**a**) and simultaneous packet transmission, SMPT, approach (**b**) which reduces higher layer packet jitter [70]

last packet of a segmented IP packet, is an additional delay experienced for this particular IP packet.

It is shown in [70] that SMPT can maintain a substantially more constant throughput for different BER values than without SMPT. TCP's congestion window is kept at a high level, not reduced due to unnecessary TCP timeouts, thus having constant high TCP throughput.

However, several problems arise with SMPT: Of course the number of available CDMA channels is limited and due to the inconstant use of channels by mobile stations an efficient channel allocation is a challenging task. This is even more difficult if multiple MS share the same channels. Furthermore, allocation of new CDMA channels might not be done immediately but needs a finite setup time which has to be taken into account in the channel allocation algorithm.

Additionally, the SNIR of the CDMA channels vary for different number of allocated channels which makes link maintenance in terms of power control and CDMA cell planning even more difficult. A possible solution would be to apply a more powerful FEC for the retransmitted packets to overcome the lower SNIR.

5.3 Packet Length for Optimum Throughput

The performance of TCP depends on many parameters, like congestion window, timeout computation, acknowledgment policy and packet length. In this section we will focus on the *packet length* as a parameter. For error-free links[5], packet length only effects the protocol overhead. However, on links with errors, the observed packet error rate depends on the packet length.

---

[5] Of course, no ARQ protocol like TCP would be necessary on error-free links. Nevertheless, this case is illustrative.

After presenting the basic context of the optimum packet length we will review the throughput efficiency on different link characteristics. Using these results we will compute the optimum packet length for maximum throughput. The results are valid for a general go-back-N ARQ protocol. A go-back-N ARQ protocol retransmits N packets when an error occurs [91]. The plain TCP protocol without any further enhancements is the most important representative of this class of ARQ protocols.

5.3.1 Basic Context

On links with errors the Block Error Ratio, BLER, depends on the Bit Error Ratio, BER, and the packet length, n. Consider an independent, identical distributed (i.i.d.) error distribution, then the BLER can be computed as [91]

$$\text{BLER} = 1 - (1 - \text{BER})^n , \tag{5.2}$$

where $1 - \text{BER}$ is the probability that a single bit is error free. $(1 - \text{BER})^n$ is the probability that a packet with length n has no bit errors. A packet is considered as faulty if a single bit within the packet is faulty.

We assume a go-back-N ARQ protocol. The overall protocol performance is given by the throughput and the delay experienced at the receiver. The throughput itself is affected by the BLER and the protocol overhead, ϱ. A packet consists of a header part with length n_H and a data part with length n_D. Then the protocol overhead, ϱ, is defined as

$$\varrho = \frac{n_\text{H}}{n_\text{D}} . \tag{5.3}$$

Because the BLER itself depends on the BER *and* the packet length, n, throughput, η, can be expressed as

$$\eta = f(\text{BLER}, \varrho) = f(\text{BER}, n, \varrho) . \tag{5.4}$$

Both parameters, protocol overhead and block error ratio, BLER, depend on the packet length. Thus, two effects occur when changing the packet length:

1. For increasing packet length, n, the BLER increases, thus throughput efficiency *decreases*.
2. For increasing packet length, n, the overhead, ϱ, decreases, thus the throughput efficiency *increases*.

Points 1 and 2 describe two different mechanisms that work against each other when changing the total packet length of the protocol. Thus some kind of *optimum packet length* can be expected.

5.3.2 Throughput Efficiency

The throughput efficiency, η, is defined as the ratio of the average number of information digits successfully accepted by the receiver per unit of time to the total number of digits that could be transmitted per unit of time. Throughput efficiency has no dimension and a value range from $0 \leq \eta \leq 1$. In the following, we will present the throughput efficiency for channels with i.i.d. errors and 2-state Markov error models.

Binary symmetric channel with error-free feedback. Let P denote the probability that the received packet of length n is accepted by the receiver. P consists of the probability that the received packet contains no errors, P_c, and the probability that a packet contains an undetectable error pattern, P_e; $P = P_c + P_e$. If we assume that $P_e \ll P_c$, then

$$P \approx P_c = 1 - \text{BLER} . \tag{5.5}$$

The throughput efficiency, η, for a go-back-N ARQ protocol operating on a forward link with i.i.d. errors and an error-free feedback link is given as [91]

$$\eta = \frac{P}{P + (1 - P)N} \cdot \frac{k}{n} . \tag{5.6}$$

N is measured in packets and it can also be considered as Round Trip Time, RTT. k denotes the number of information bits per packet of length n. The RTT (N) becomes significant when the channel error rate, $\text{BLER} = 1 - P$, becomes high. The term k/n is originally the code rate of a (n, k) linear code for FEC. But if we assume no FEC, as it is in the TCP/IP protocol, we can use this term for modeling the protocol overhead, ϱ. Thus the code rate k/n can be expressed as

$$\frac{k}{n} = \frac{n_D}{n} = \frac{n_D}{n_H + n_D} = \frac{1}{1 + \varrho} . \tag{5.7}$$

Finally, the throughput efficiency can be expressed as desired in (5.4) by substituting (5.5, 5.2 and 5.7) into (5.6). This results in

$$\eta = \frac{(1 - \text{BER})^{n_H + n_D}}{(1 - \text{BER})^{n_H + n_D}(1 - N) + N} \cdot \frac{n_D}{n_H + n_D} . \tag{5.8}$$

Equivalently, the throughput efficiency can be expressed as a function of overhead, ϱ,

$$\eta = \frac{(1 - \text{BER})^{n_H(1 + 1/\varrho)}}{(1 - \text{BER})^{n_H(1 + 1/\varrho)}(1 - N) + N} \cdot \frac{1}{1 + \varrho} . \tag{5.9}$$

A numerical evaluation of (5.8) is plotted in Fig. 5.8. The header length is constant, $n_H = 40$ bytes, which corresponds to the standard TCP/IP header without compression. The Round Trip Time was chosen as RTT = 4 packets. Figure 5.8(a) shows the throughput efficiency as a function of BER and the data packet length, n_D. Figure 5.8(b) is a contour plot of iso-efficiency lines for throughput values ranging between $\eta = 0.1$–0.9. Every curve represents an iso-efficiency line (contour line of Fig. 5.8(a)) for a specific constant throughput efficiency, η. As expected, throughput efficiency increases with decreasing

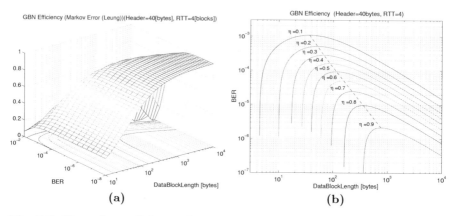

Fig. 5.8. Throughput efficiency of an erroneous forward link with i.i.d. error characteristic and with an ideal feedback link. The header length , n_H, is 40 bytes and the round trip time, RTT, is 4 packets. Plot **(b)** shows the contour lines of the three-dimensional plot **(a)** for throughput efficiencies 0.1, 0.2, ..., 0.9

BER. But it is *decreasing* when the data block length, n_D, increases, at least if n_D is above a certain threshold value.

Taking a closer look at (5.9), the throughput efficiency will diminish as $1/\varrho$ (driven by the second term) in the limit of small blocks ($\varrho \geq 1$). For large data block size, n_D, (i.e. $\varrho \ll 1$) the BLER is raised with the power of the increasing block sizes, at constant BER. Thus the throughput efficiency decreases with decreasing overhead.

Markov errors with error-free feedback. Consider a Markov block error model with two states, similar to that presented in Sect. 3.4, (3.8). The transition probability matrix has the general form,

$$\mathbf{A} = \begin{bmatrix} A_{00} & A_{01} \\ A_{10} & A_{11} \end{bmatrix} . \tag{5.10}$$

According to the definitions in [90], a *clustering coefficient*, γ, can be defined as,

$$\gamma = A_{00} + A_{11}, \quad 0 \leq \gamma \leq 2 . \tag{5.11}$$

This indicates the probability of clustering of transmissions in error, A_{11}, and the probability of clustering of error-free transmission periods, A_{00}. A high value of the clustering coefficient, $\gamma > 1$, indicates high correlation between two consecutive blocks (error-free or packets in error). Whereas low values of γ mark more statistical independence between two consecutive blocks.

The throughput efficiency for a go-back-N ARQ protocol under a two-state Markov error model with state transition matrix, $\mathbf{A}$, in the forward link and error-free feedback link is presented in [90]. Accordingly, the throughput efficiency can be expressed in general as

$$\eta = \frac{\mathrm{E}\{K\}}{N + \mathrm{E}\{K\}} \quad . \tag{5.12}$$

K denotes the number of accepted blocks in a single cycle, $\mathrm{E}\{K\}$ correspondingly the average of K. A cycle starts with a block error and ends with the first block error after the Nth block following the first block in error. With the general Markov error model (5.10) the throughput efficiency can be expressed by [90]

$$\eta'_{\mathrm{GBN},Mk} = \frac{A_{10}(1 - (\gamma - 1)^N)}{A_{10}(1 - (\gamma - 1)^N) + N(2 - \gamma)(2 - \gamma - A_{10})} \quad . \tag{5.13}$$

To contribute the protocol overhead an overhead term must also be added. The total throughput efficiency is thus

$$\eta_{\mathrm{GBN},Mk} = \frac{A_{10}(1 - (\gamma - 1)^N)}{A_{10}(1 - (\gamma - 1)^N) + N(2 - \gamma)(2 - \gamma - A_{10})} \cdot \frac{n_{\mathrm{D}}}{n_{\mathrm{H}} + n_{\mathrm{D}}} \quad . \tag{5.14}$$

Of course, all Markov parameters depend on the block length, n, therefore $\mathbf{A} = \mathbf{A}(n)$.

A numerical evaluation of (5.14) is plotted in Fig. 5.9. We estimated the Markov parameters using sample model parameters for a GSM data traffic channel TCH/F9.6. The error sequences have been generated with the link level simulation environment already presented in Sect. 3.3. We used the 'Typical Urban' channel model and a mobile station velocity of 10 km/h. The carrier-to-interference ratio, C/I, has been varied between 0 dB and 20 dB.

Qualitative behavior of the modeled throughput is essentially the same as for the i.i.d. error model (compare with Fig. 5.8). For small block lengths the throughput increases with increasing block length, but for large block lengths throughput decreases with increasing block length. Throughput diminishes with decreasing C/I. The quantitative behavior can not be compared exactly, because of the different scale on the BER and C/I axis.

Markov errors on forward and feedback link. Investigations of the throughput efficiency on links with Markov error patterns on both forward and feedback link have been presented in [85]. There, independent two-state Markov models for each link are assumed. Basically the authors combine the transition matrices of the Markov models (5.10) for forward and feedback link into a combined transition matrix with four states. Let f denote the index

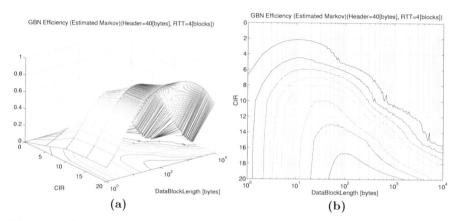

Fig. 5.9. Throughput efficiency from (5.14) with estimated Markov parameters for GSM TCH/F9.6 for an ideal feedback link. The plot on the right-hand side **(b)** shows the contour lines of the 3D plot **(a)**, i.e. iso-efficiency lines

for the forward channel and b for the backward channel. Then the combined transition matrix, $\mathbf{A}_{fb}$, can be obtained as

$$\mathbf{A}_{fb} = \begin{bmatrix} A_{00,f}A_{00,b} & A_{00,f}A_{01,b} & A_{01,f}A_{00,b} & A_{01,f}A_{01,b} \\ A_{00,f}A_{10,b} & A_{00,f}A_{11,b} & A_{01,f}A_{10,b} & A_{01,f}A_{11,b} \\ A_{10,f}A_{00,b} & A_{10,f}A_{01,b} & A_{11,f}A_{00,b} & A_{11,f}A_{01,b} \\ A_{10,f}A_{10,b} & A_{10,f}A_{11,b} & A_{11,f}A_{10,b} & A_{11,f}A_{11,b} \end{bmatrix} \equiv \begin{bmatrix} \alpha_{11} & \alpha_{12} & \alpha_{13} & \alpha_{14} \\ \alpha_{21} & \alpha_{22} & \alpha_{23} & \alpha_{24} \\ \alpha_{31} & \alpha_{32} & \alpha_{33} & \alpha_{34} \\ \alpha_{41} & \alpha_{42} & \alpha_{43} & \alpha_{44} \end{bmatrix}.$$

$$(5.15)$$

Additionally the N-step transition matrix, $\mathbf{A}_N \equiv \mathbf{A}_{fb}^N$, representing the transition probabilities between blocks that are N-blocks apart from each other, is defined as

$$\mathbf{A}_N = \begin{bmatrix} e_{11} & e_{12} & e_{13} & e_{14} \\ e_{21} & e_{22} & e_{23} & e_{24} \\ e_{31} & e_{32} & e_{33} & e_{34} \\ e_{41} & e_{42} & e_{43} & e_{44} \end{bmatrix}. \qquad (5.16)$$

Throughput efficiency is again determined by (5.12), but in this 'erroneous feedback link' case, $E\{K\}$, the average number of accepted blocks during a single cycle, is computed from

$$E\{K\}_{Mk,\text{back}} = \frac{\alpha_{12}e21 + \alpha_{13}e31 + \alpha_{14}e31}{(1 - \alpha_{11})^2}. \qquad (5.17)$$

When the backward channel is error free, (5.17) reduces to $E\{K\} = e_{31}/\alpha_{12}$ which leads to (5.13).

To contribute the protocol overhead, the overhead term $n_\mathrm{D}/(n_\mathrm{H} + n_\mathrm{D})$ must be added. Then the total throughput results in

$$\eta_{\mathrm{GBN},Mk,\mathrm{back}} = \frac{E\{K\}_{Mk,\mathrm{back}}}{N + E\{K\}_{Mk,\mathrm{back}}} \cdot \frac{n_{\mathrm{D}}}{n_{\mathrm{H}} + n_{\mathrm{D}}} \ . \tag{5.18}$$

Numerical evaluations of (5.18) for Markov parameters estimated from the GSM TCH/F9.6 traffic channel (COST 207 channel model at $10\,\mathrm{km/h}$) are plotted in Fig. 5.10. General behavior of the throughput efficiency has not

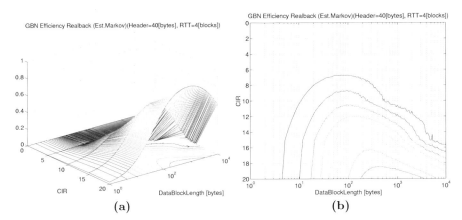

Fig. 5.10. Throughput efficiency from (5.18) with estimated Markov parameters for GSM TCH/F9.6 for a Markov error model for both forward and feedback link. The plot on the right-hand side **(b)** shows the contour lines of the 3D plot **(a)**

changed compared to the previous cases. However, the throughput efficiency is, in general, lower for the entire C/I range. Additionally, the maximum throughput values have been shifted towards *higher* block lengths!

5.3.3 Optimum Packet Length

As already outlined in Sect. 5.3.1, an optimum packet length should exist for each specific value of BER. It is defined as

$$n_{\mathrm{D,opt}}(N, n_{\mathrm{H}}, \mathrm{BER}) = \max_{n_{\mathrm{D}}} \eta(N, n_{\mathrm{D}}, n_{\mathrm{H}}, \mathrm{BER}) \ . \tag{5.19}$$

In the following we will compute the optimum data block length for the three different error-statistics cases presented in the previous section.

Binary symmetric channel with error-free feedback. Equation 5.8 defines the throughput efficiency on a link with i.i.d error pattern on the forward link only. Solving (5.8) for BER as a function of η, N, n_{H} and n_{D}, yields

$$\mathrm{BER} = 1 - \left(\frac{\eta N(n_{\mathrm{D}} + n_{\mathrm{H}})}{n_{\mathrm{D}}(1 + \eta(N-1)) + \eta n_{\mathrm{H}}(N-1)} \right)^{\frac{1}{n_{\mathrm{D}} + n_{\mathrm{H}}}} \ . \tag{5.20}$$

The optimum data block size is the solution of

$$\frac{\partial \mathrm{BER}(\eta, N, n_\mathrm{D}, n_\mathrm{H})}{\partial n_\mathrm{D}} = 0 \ , \tag{5.21}$$

for n_D, i.e. the maximum of the BER() function (5.20). Unfortunately, there exists no exact analytical solution.

A numerical evaluation of the throughput efficiency (5.20) for a header length of $n_\mathrm{H} = 40$ bytes and a round trip time RTT $= 4$ blocks is already plotted in Fig. 5.8(b). The figure shows iso-efficiency lines, connecting pairs of BER and data block length values, n_D, that lead to the same throughput efficiency.

For a specific BER (e.g. the mean BER on the channel, BER $\approx 10^{-5}$), there exists an *optimum* data block length ($n_\mathrm{D} \approx 350$ bytes) where the highest throughput efficiency (here $\eta = 0.8$) is obtained. The dashed line in Fig. 5.8(b) connects all the (BER,n_D) pairs that give *optimum* efficiency. It can be conveniently used to determine the optimum data block length when a specific BER is observed on the transmission link.

The optimum (BER,n_D) pairs for highest efficiency for various header block lengths and round trip times are plotted in Fig. 5.11.

A short example illustrates the usage of this curves: Consider the TCP/IP protocol with standard header length of $n_\mathrm{H} = 40$ bytes. If the mean link BER is BER $\approx 10^{-4}$ in the forward direction, and the mean round trip time is RTT $= 4$ packets then we read from Fig. 5.11 that the data packet size for maximum throughput efficiency should be $n_\mathrm{D} \approx 120$ bytes.

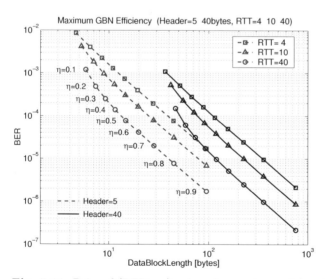

Fig. 5.11. Pairs of (BER,n_D) values for maximum throughput efficiency of a forward link with i.i.d. error characteristic and ideal feedback link

In general, with constant header length the data block length should be increased with decreasing BER and vice versa. For a header length of 40 bytes (TCP/IP) and a RTT of 4 blocks the data block length should be chosen as 100 to 1000 bytes for BER $\approx 10^{-4}$–10^{-6}. Thus *adaptive* data block length adjustment dependent on the current mean BER would lead to optimum system throughput efficiency.

Figure 5.11 also illustrates the strong impact of the header length on the optimum throughput efficiency. To achieve maximum efficiency, the block length should be chosen in line with header length. Conversely, an increase in header length should always be accompanied by an increase in data block length.

Markov errors with error-free feedback. The throughput efficiency for Markov errors with error-free feedback is given by (5.14). Because no explicit dependency on BER is available, an analytical solution is again not possible. But a numerical evaluation is still feasible.

Looking at Fig. 5.9, the optimum data packet length, n_D, for a specific throughput efficiency can be determined by finding the minimum C/I value for this specific iso-efficiency line. In this way, $(C/I, n_D)$ pairs with optimum efficiency can be extracted. Doing this for various different parameters, we obtain Fig. 5.12.

General dependencies between C/I (which corresponds to BER in the previous case) and n_D remain, better C/I corresponds to higher optimum block length. The optimum data block length depends more heavily on the header length, n_H, than on the round trip time, RTT. Thus the observed

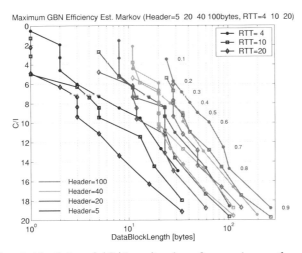

Fig. 5.12. Pairs of $(C/I, n_D)$ values for maximum throughput efficiency for an erroneous forward link under Markov error structure and ideal feedback link. The Markov parameters are estimated for GSM TCH/F9.6 traffic channel and COST 207, typical urban channel model with a mobile station velocity of 10 km/h

RTT is not critical. For moderate to high C/I values ($10\,\mathrm{dB} \leq C/I \leq 20\,\mathrm{dB}$) and a standard TCP/IP header length of $n_\mathrm{D} = 40\,\mathrm{bytes}$, data block length should be selected as $n_\mathrm{D} \approx 20\text{--}200\,\mathrm{bytes}$. The lower values for lower C/I and higher values for higher C/I.

Markov errors on forward and feedback link. Doing the same procedure as in the previous paragraph for the link with Markov errors on forward and backward channels, we obtain again $(C/I, n_\mathrm{D})$ pairs for optimum efficiency. The results for various parameter settings are plotted in Fig. 5.13.

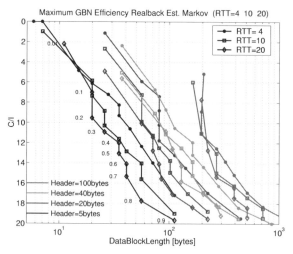

Fig. 5.13. Pairs of $(C/I, n_\mathrm{D})$ values for maximum throughput efficiency for both erroneous forward and feedback link under Markov error structure. The Markov parameters are estimated for GSM TCH/F9.6 traffic channel and COST 207, typical urban channel model with a mobile station velocity of $10\,\mathrm{km/h}$

Quantifying the example of a TCP/IP data transmission without header compression ($n_\mathrm{H} = 40\,\mathrm{bytes}$) and a round trip time of RTT=4 packets, a data packet length, n_D, in the range from approximately 200 to 1000 bytes should be used for a C/I range from 15 to 20 dB. The higher the C/I value, the longer the data packets should be. With TCP/IP header compression activated ($n_\mathrm{H} \approx 5\,\mathrm{bytes}$), the optimum data block length reduces to 50 to 120 bytes for the same C/I range.

Comparing this result with that for the error-free feedback link, shifted maximum throughput values to higher block lengths can be observed. This is a rather interesting result, because feedback packets are always modeled as packets with length of a header of the forward link, n_H. Thus, if acknowledgment packets can get lost too, the forward link data packets should be longer.

5.3.4 Conclusions

We presented the throughput efficiency for a go-back-N ARQ protocol over three different error characteristics on the link. These are, namely, the i.i.d. error pattern and the general 2-state Markov model on the forward link only and 2-state Markov error models on forward and feedback link. Because throughput efficiency depends in a conflicting manner on overhead and on BER when changing the block length, an *optimum* block length does exist.

In general, the data block length should be increased for decreasing BER ($\equiv$ increasing C/I). This is the main dependency. Additionally the overhead, ϱ, and the round trip time, RTT, influence the optimum block length. If larger headers are used in every block, longer data block lengths are also required for maximum throughput. RTT is only a minor factor considering the optimum block length.

A summary of all parameter dependencies is listed in Table 5.2.

Table 5.2. Qualitative dependencies of relevant parameters to achieve maximum throughput efficiency with the optimum block length, $n_{D,opt}$

Parameter behavior		Optimum block length dependency	
BER	$\downarrow$	n_D	$\Uparrow$
C/I	$\uparrow$	n_D	$\Uparrow$
n_H	$\uparrow$	n_D	$\Uparrow$
RTT	$\uparrow$	n_D	$\downarrow$

5.4 TCP over GSM-CSD, Simulations and Measurements

Because of the complex algorithms included in the TCP protocol, it is almost impossible to predict the performance of TCP in real environments. This is especially true in highly unpredictable environments like the mobile radio channel. High and varying BER and also frequent handovers impair the performance of TCP used over a mobile radio link.

To obtain more realistic projections about the performance of TCP used over GSM, we performed both simulations *and* measurements. Naturally, simulation of complex systems require specific assumptions to restrict the effort to a manageable extent. However, the insight gained by having access to internal system parameters and the possibility of tuning *all* system parameters make simulation results nevertheless valuable. The measurements have been conducted in various networks in Austria.

All simulations have been done with 'Ptolemy', which is a heterogenous, object-oriented simulation tool, developed by the University of Berkeley [78].

Because of the differences in GSM-CSD and GSM-GPRS, we investigate both systems separately. GPRS simulations will be shown in Sect. 5.5.

5.4.1 Simulation Model for GSM-CSD

For simulation of TCP/IP on GSM circuit-switched data services we used a standard scenario: A user wants to download data from a host located at the fixed network over a GSM CSD traffic channel. The user dials in at an *Internet Service Provider* (ISP) which provides access to the Internet. The call is routed through the GSM mobile network to the *Interworking Function* (IWF). The IWF sets up a modem connection to the ISP. In our example we assume that data is downloaded directly from the ISP's host. There is no further connection to the Internet assumed. The overall structure of the model is shown in Fig. 5.14.

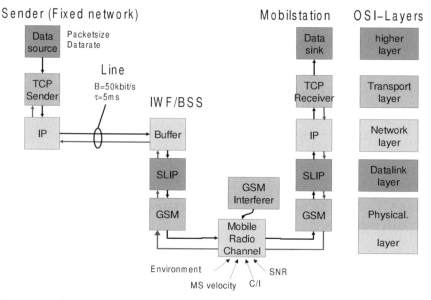

Fig. 5.14. Overall structure of the model for simulation of TCP/IP download data transmission over a GSM circuit-switched traffic channel

The sender is located at the fixed network. A data source offers data packets with constant payload length and constant rate to the TCP sender (offered load). This emulates FTP or bulk data transfer.

The TCP sender encapsulates the data packets into TCP packets with 20 bytes header overhead. Then the TCP packets are sent to the IP layer using the TCP protocol mechanisms. All relevant mechanisms like slow start, congestion avoidance, sliding window technique, fast retransmission, fast recovery, RTT estimation and adaptive retransmission timeout have been implemented.

Receiving TCP packets, the IP layer adds the 20 byte long standard IP header. Then IP packets pass a transmission line with a bandwidth of $B = 50\,\text{kbit/s}$ and a delay of $\tau = 5\,\text{ms}$. The transmission line should simulate a modem connection from the fixed host to the interworking function at the MSC. At the MSC the data is buffered when entering the mobile network. Afterwards the IP packets are framed into SLIP packets and sent to the GSM base station subsystem (BSS). This includes, finally, the GSM base station.

This model is somewhat inaccurate: SLIP is a data link layer protocol and performs a point-to-point connection. However, the point-to-point connection should start at the beginning of the modem connection (in this model at the TCP sender at the ISP) and end at the mobile station. Here the SLIP connection is modeled only over the GSM link. Thus SLIP protocol overhead is not included in the packets traveling over the modem line, which is the case in reality. But as the simulated modem line has a bandwidth of $B = 50\,\text{kbit/s}$ and the GSM link at maximum a bandwidth of $14.4\,\text{kbit/s}$, the GSM link is the bottleneck of the connection. Thus, it is not that important to model the connections outside of the mobile network exactly, but rather the GSM link itself and the protocols running over the GSM link. All included protocol overheads burden the GSM link anyway, thus the model should be sufficiently realistic concerning this detail.

The GSM link is modeled by the link level simulation model presented in Sect. 3.3. The main parameters for this GSM link level model including the mobile radio channel and a GSM interferer are: Environment ('Typical Urban', ...), mobile station velocity, SNR and C/I.

At the mobile station the packets travel the entire chain of different protocol layers upwards. Every protocol decapsulates the corresponding upper layer protocols. The IP layer additionally performs reassembling of the fragmented packet and a check sum test. Arriving uncorrupted at the TCP receiver, an acknowledgment packet is generated, according to the TCP protocol rules. If the packet has been received correctly and in order, it is handled upwards to the data sink which ends the individual packet transmission. But it is ended at the TCP sender only if it is subsequently correctly acknowledged at the TCP sender. Therefore the acknowledgment packet is sent all the way back through all protocol layers to the TCP sender. Arriving there, it definitely ends the individual packet transfer. Additionally, it triggers the TCP sender to send new TCP packets according to the TCP rules.

The default simulation parameters are summarized in Table 5.3. Some parameter values imply a specific behavior of the simulation environment which we will point out in the following:

Table 5.3. Default values of the most important simulation parameters for both TCH/F9.6 and TCH/F14.4 traffic channel

Parameter	Default value	
	TCH/F9.6	TCH/F14.4
Data source		
Packet size	500 bytes	
Offered load	12 kbit/s	17.6 kbit/s
TCP		
Version	Tahoe	
Header length	20 bytes	
IP		
Maximum IP packet size	65 536	
Header length	20 bytes	
SLIP		
Overhead	≥2 bytes	
IWF		
Buffer size	∞	
GSM physical layer		
Mode	Transparent	
Environment	'Typical urban'	
Traffic channel	TCH/F9.6	TCH/F14.4
MS velocity	Variable	
C/I	Variable	
SNR	100 dB ($\Rightarrow$ Interference limited)	
Forward channel	COST 207 model	
Backward channel	COST 207 model	

Offered load is about 25% above the maximum possible throughput over the GSM radio interface for user data. Thus there is always enough data available for sending.

Maximum IP packet size is 65 536. Because data *packet size* is always below *Maximum IP packet size*, fragmentation of TCP packets never occurs.

IWF buffer size is infinity: We assume there is enough buffer space available. Thus no packet loss because of buffer overflow is possible. The only source of errors is the GSM radio interface.

SLIP overhead is *not* deterministic. SLIP replaces each END and ESC characters in the payload by a special two-byte sequence. As random payload data is generated by the data source, the number of END and ESC replacements is also random.

We used the transparent GSM traffic channels, i.e. no RLP protocol.

5.4.2 Simulation Results for GSM-CSD

We will now present simulation results obtained for standard GSM circuit-switched data channels. Both TCH/F9.6 and TCH/F14.4 traffic channels

have been considered. Remember that all CSD simulations have been performed for *transparent* data transmission. Only TCP retransmissions are responsible for reliable data transfer.

As the main parameters for the end user are net throughput and end-to-
end delay, we concentrate on these.

Net throughput. The throughput presented here is the net data throughput on the TCP protocol level. An application using the TCP protocol as
transport protocol can expect this throughput.

The dependence of the average net throughput on the C/I is plotted in
Fig. 5.15. Both traffic channels, TCH/F9.6 and TCH/F14.4, are shown for a

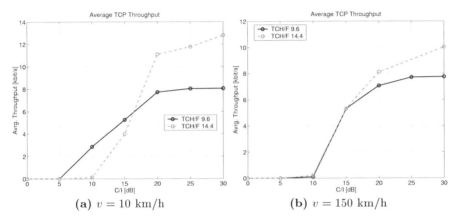

(a) $v = 10$ km/h   **(b)** $v = 150$ km/h

Fig. 5.15. Average net throughput versus C/I on the radio channel: **(a)** corresponds to a mobile station velocity of $v = 10$ km/h and **(b)** to $v = 150$ km/h. GSM
transparent mode was used.

mobile station speed of $v = 10$ km/h in Fig. 5.15(a) and for $v = 150$ km/h in
Fig. 5.15(b).

Looking at Fig. 5.15(a) (10 km/h) a maximum throughput of about
8 kbit/s for the TCH/F9.6 channel is observed for $C/I \geq 20$ dB. Measurable
throughput values start at $C/I \geq 5$ dB and increase linearly with C/I. For
the 14.4-kbit/s traffic channel a maximum throughput of almost 13 kbit/s
at $C/I = 30$ dB has been obtained. However, utilizable data transmission
only starts beyond $C/I \geq 10$ dB. But 84% of the maximum throughput of
13 kbit/s are already reached at $C/I = 20$ dB.

Thus two observations are important when comparing 9.6 and 14.4-
kbit/s traffic channels: At first, TCH/F9.6 has a lower C/I threshold value
for achieving usable throughput of $C/I > 5$ dB whereas for TCH/F14.4
at least $C/I > 10$ dB are necessary. But secondly, TCH/F14.4 shows a
steeper increase in throughput when increasing C/I than TCH/F9.6. Thus
the radio channel quality (C/I) has a stronger impact on TCH/F14.4 than

for TCH/F9.6, which is twofold: TCH/F14.4 needs a better C/I of about 5 dB, and it is more susceptible to fades where C/I is below its mean (see Fig. 5.15(a)). This behavior reflects the worse channel coding for TCH/F14.4.

Different behavior can be seen in Fig. 5.15(b), which are the simulation results for a mobile station velocity of $v = 150$ km/h. Now both traffic channels need a C/I above $C/I > 10$ dB which is the *common* threshold C/I. The maximum throughput for TCH/F9.6 reduces to about 7.8 kbit/s and to some 10 kbit/s for TCH/F14.4. Thus throughput degradation for high mobile station velocities is only 10% for TCH/F9.6, whereas about 24% degradation can be observed for TCH/F14.4.

Net throughput as a function of TCP packet size is plotted in Fig. 5.16. Throughput for TCH/F9.6, $C/I = 20$ dB and both velocities $v = 10$ km/h

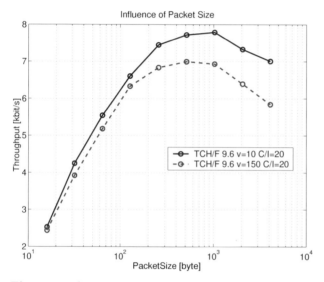

Fig. 5.16. Average net throughput versus TCP packet size for TCH/F9.6 and $C/I = 20$ dB

and $v = 150$ km/h is shown. Clearly a maximum for both curves can be observed. The shape of the curves proves the behavior expected from Sect. 5.3, where an optimum block length was calculated. The optimum block length for maximum throughput for TCH/F9.6 at $v = 10$ km/h is $n_{D,opt} \approx 1000$ bytes. It reduces to $n_{D,opt} \approx 250$ bytes at $v = 150$ km/h.

These results correspond very well to the theoretical ones presented in Fig. 5.13 for a header length of 40 bytes (TCP/IP header) and $C/I = 20$ dB. As already explained in Sect. 5.3, the throughput below the optimum block length is mainly determined by the protocol overhead. Above the optimum block length the increasing block error rate for increasing block length is responsible for reduced throughput.

End-to-end delay. A second performance measure is the delay, τ. What is presented here is the *end-to-end* delay a TCP user (i.e. the application, the protocol level above TCP) experiences. Figure 5.17 shows the mean end-to-end delay for TCH/F9.6 and TCH/F14.4 as a function of C/I. Remember that the delay definition in GPRS and UMTS is different. There, only the delay within the mobile network is included. The delay investigated in this section also includes the external network delay from the IWF to the fixed host.

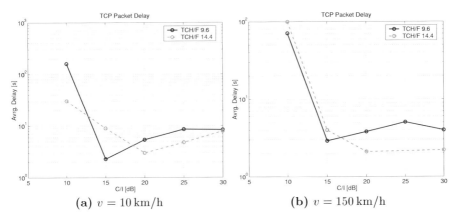

Fig. 5.17. Average end-to-end delay, τ, versus C/I: **(a)** corresponds to a mobile station velocity of $v = 10\,\text{km/h}$ and **(b)** to $v = 150\,\text{km/h}$

In the left-hand side plot (Fig. 5.17(a)) a mobile station velocity of $v = 10\,\text{km/h}$ was assumed. For TCH/F9.6 and poor radio channel ($C/I < 15\,\text{dB}$) the delay increases to $\tau > 100\,\text{s}$, which is unacceptable as QoS of a circuit-switched connection. As C/I increases, the mean delay decreases to a minimum ($\tau_{\min,9.6} \approx 2.3\,\text{s}$) at $C/I = 15\,\text{dB}$ to increase again for $C/I > 15\,\text{dB}$. The same shape can be observed for TCH/F14.4, but the minimum delay ($\tau_{\min,14.4} \approx 3\,\text{s}$) is shifted to $C/I \approx 20\,\text{dB}$.

In principle the same behavior is found in Fig. 5.17(b), where a mobile station velocity of $v = 150\,\text{km/h}$ was set. Again a very high delay for $C/I < 10\,\text{dB}$ of about $\tau \approx 100\,\text{s}$ is seen. Also τ increases after reaching a minimum delay for increasing C/I. Remarkably, the minimum delays are again at $C/I = 15\,\text{dB}$ (TCH F/9.6) and $C/I = 10\,\text{dB}$ (TCH/F14.4).

The special shape of the mean delay curves in Fig. 5.17 relies on two facts. First, the poor radio channel is responsible for the large delay for small C/I, i.e. many errors require frequent retransmissions. However, when errors start to diminish (increasing C/I), the influence of the IWF buffer is more pronounced. This can be verified with Fig. 5.18. Figure 5.18 shows the average IWF buffer queuing size in dependency of C/I. For low C/I there is almost no buffer space used, the mean queuing size is zero or almost zero. But the

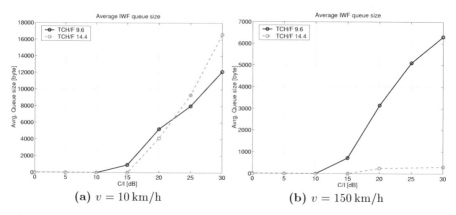

Fig. 5.18. Average IWF buffer queue size versus C/I for both TCH/F9.6 and TCH/F14.4. The mobile station velocity was set to $v = 10$ km/h **(a)** and $v = 150$ km/h **(b)**

mean queuing size rises with increasing C/I. If only a few errors occur on the radio interface, TCP increases its congestion window linearly to the maximum congestion window. Note that the IWF buffer size was infinite, so no packets got lost due to buffer overflow. Thus TCP fills the buffer more and more, the number of packets waiting for transmission in the buffer starts to increase. The more packets that are in the queue, the longer a packet needs to get finally transmitted over the radio path. Also, TCP's reaction time increases because at least all packets in the buffer lie in between a possible wrong packet, and the packet where TCP reacts upon the erroneous packet. This is the explanation for the shape of the mean delay curves in Fig. 5.17.

The difference between Fig. 5.18(a) and Fig. 5.18(b) is the mobile station velocity. For TCH/F9.6 the behavior is almost identical for $v = 10$ km/h and $v = 150$ km/h. Buffer queue size increases for higher C/I. The absolute values differ by about 40%. Looking at TCH/F14.4, a much larger difference from $v = 10$ km/h to $v = 150$ km/h can be found. This corresponds to the observation from Fig. 5.15. As channel coding is much worse for TCH/F14.4, a higher velocity (worse radio channel) has a much higher impact on performance. Thus the mean occupied buffer space is much smaller for $v = 150$ km/h than for $v = 10$ km/h. The link quality is worse for $v = 150$ km/h, thus the TCP packet injection rate is smaller which leads to smaller buffer occupancy. A stable state is not reached.

The corresponding TCP packet delay distributions for the mean values in Fig. 5.17(a) are plotted in Fig. 5.19. The left-hand side plot (Fig. 5.19(a)) shows the delay distributions for TCH/F9.6. The distributions are almost symmetrical around the mean values, i.e. smaller delays have almost the same probability as longer delay values. Exceptions to this mark the very small and very large delays. Delays below ≈ 0.7 s do not exist. A minimum delay

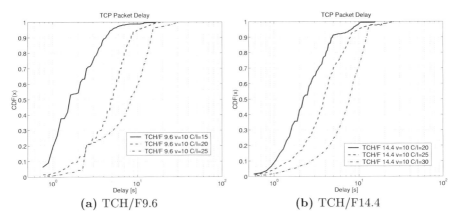

Fig. 5.19. CDF of TCP packet delay for a MS velocity of $v = 10\,\text{km/h}$. **(a)** shows delay distribution for TCH/F9.6 and **(b)** for TCH/F14.4

is already consumed by the pure transmission time ($> 450\,\text{ms}$ for TCH/F9.6 and 500 bytes TCP packets) and the additional queuing delays in the other components, especially IWF. The upper end of the CDFs correspond to very long delays. Although the radio channel is not too bad at a $C/I = 25\,\text{dB}$, delays $\tau > 20\,\text{s}$ occur in about 3% of all packets.

The right-hand side plot shows delay distributions for TCH/F14.4 (Fig. 5.19(b)). The distribution functions are almost equal to the previous case, only the mean value is shifted accordingly. However, very long delays occur more often than for TCH F/9.6. Also, because the minimum delay is smaller for TCH/F14.4 ($> 300\,\text{ms}$ for TCP data packet length of 500 bytes) the lower left end is shifted to smaller delays. Despite this, the probability for smaller delays has not changed significantly.

End-to-end delay vs. TCP packet size. The influence of TCP packet size on the mean end-to-end delay is plotted in Fig. 5.20. Only TCH/F9.6 with $v = 10\,\text{km/h}$ and $v - 150\,\text{km/h}$ mobile station velocity is shown. In general, τ increases with larger packet sizes. This results from the following: Firstly, the transmission time itself increases with packet size. For a 4096-byte data packet, pure transmission time is approximately 3.45 s, which is already about 24% of the total delay. Secondly, the BLER increases for higher block sizes and yet constant C/I. Thus, more packets also have errors and consequently more retransmissions occur. For these reasons, the mean delay increases from a minimum of $\tau = 2.5\,\text{s}$ (TCH/F9.6, $v = 10\,\text{km/h}$) or $\tau = 1.8\,\text{s}$ (TCH/F9.6, $v = 150\,\text{km/h}$) for a packet size of 32 bytes to about 14.5 s ($v = 10\,\text{km/h}$) and 13 s ($v = 150\,\text{km/h}$) for a packet size of 4096 bytes. Thus, if packet delay matters in the application (i.e. interactive traffic), packet size should be small.

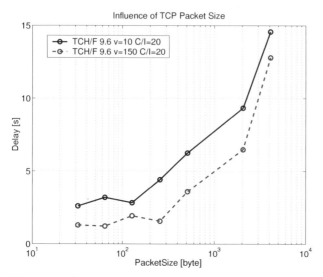

Fig. 5.20. Average packet delay versus TCP packet size for TCH/F9.6. The mobile station velocity was $v = 10\,\mathrm{km/h}$ and $v = 150\,\mathrm{km/h}$, and $C/I = 20\,\mathrm{dB}$

5.4.3 Measurement Setup for GSM-CSD

For both verification of simulation results and to obtain real-world results, we performed measurements of the GSM circuit-switched traffic channels. The original GSM 9.6-kbit/s traffic channel (transparent and non-transparent) as well as GSM Phase 2+ traffic channels like TCH/F14.4 and HSCSD with 28.8 kbit/s have been measured. All measurements were conducted in various networks in Austria.

We measured *downlink* data transfer only, i.e. data transfer from the data server to the mobile terminal. All presented results are mean values observed from several independent measurements.

We used two different measurement setups in the following, called Setup A and Setup B. The basic measurement setup was identical for both. At the mobile station side only minor differences exist. We will explain Setup A in detail in the beginning and only point out the differences to Setup B in the second part of this section.

Measurement Setup A. Setup A is illustrated in Fig. 5.21. The GSM mobile station (MS) was connected via a serial line to a mobile terminal (Laptop). A dial-up connection was established between the MS and an Internet Service Provider (ISP). This connection consists of several parts: The most important one in the entire data path is the radio interface, which is also the bottleneck link. It is terminated at the base station. Then the internal network of the mobile network operator acts as as intermediate part until the interworking function (IWF). At the IWF a modem is used to modulate the data for

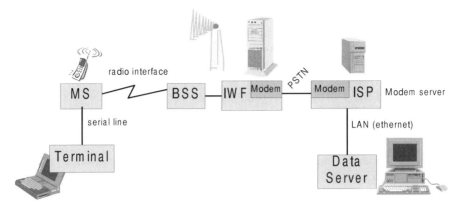

Fig. 5.21. Measurement Setup A for data transmission over GSM CSD traffic channels

transmission over the *Public Switched Telephone Network* (PSTN). At the ISP again a modem terminates the dial-up connection path. Within the ISP a Local Area Network (LAN) is used to connect the modem server to the actual data server. The LAN-type was Ethernet, which implies a maximum TCP packet size of 1500 bytes.

The corresponding protocol stack for this measurement setup is shown in Fig. 5.22. TCP was used as end-to-end transport protocol. TCP is only active in both endpoints. It uses IP as network layer protocol. In the ISP internal network IP also maintains the routing capabilities of the LAN, thus IP is also active in the modem server at the ISP. Spanning from the modem server at the ISP to the mobile terminal, the link level protocol provides IP packet transmission between these two endpoints. As link layer protocols, both SLIP and PPP had been used. Between ISP modem server and IWF

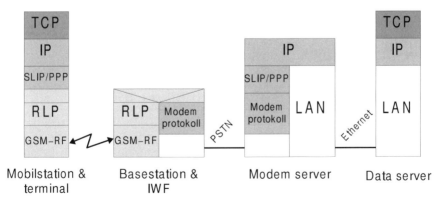

Fig. 5.22. Corresponding protocol stack for the measurement Setup A from Fig. 5.21

a special modem protocol maintained the connection over the PSTN. No explicit control over this protocol was possible. Finally, the GSM network is responsible for the connection from the mobile station to the IWF. In non-transparent traffic channels, the RLP protocol is active between the MS and the IWF, in transparent mode, RLP is not applied.

On top of the TCP layer, in principle, any application layer protocol can be utilized. In all measurements we used the File Transfer Protocol (FTP) as application layer protocol. With the FTP connection a file transfer with large file size (400 kbyte) has been established. The transferred test data files consisted of compressed data to avoid falsification due to the modem protocol between IWF and ISP modem server. This is because the modem protocol itself may have data compression activated.

To measure the TCP parameters we used a special tool, `tcpdump` [82]. Utilizing this, all relevant parameters, like packet size, throughput, retransmissions and timing values can be analyzed. The *Maximum Transmission Unit* (MTU) from the link level protocol (SLIP or PPP) was an input parameter and set at link establishment. Since the LAN connection in the internal network of the ISP uses Ethernet, the maximum path MTU[6] was in either case limited to 1500 bytes, the Ethernet limit. However, if the MTU of the link level (PPP, SLIP) was set below 1500 bytes, the TCP segment size also shows the same segment size (with TCP/IP header). Thus the setting of the link level path MTU was used by TCP to set the TCP segment size. This is standard behavior of TCP. With this feature fragmentation of packets is also avoided. We used this feature to set the TCP packet size explicitly. We varied it from 110 up to 1460 bytes.

Table 5.4 summarizes the measurement parameters.

Table 5.4. Default values of parameters for GSM CSD measurements

Application layer protocol	FTP
Transfer file size	300 . . . 500 kbytes
TCP/IP header compression	ON or OFF
Link layer MTU	varied: 150 . . . 1500 bytes
(⇒TCP segment size)	110 . . . 1460 bytes
Link layer protocol	SLIP, PPP
GSM mode	transparent, non-transparent
GSM traffic channel	TCH/F9.6, TCH/F14.4, HSCSD 28.8

Measurement Setup B. A slightly different measurement setup, Setup B, was used to get additional GSM specific parameters during a data transfer. Setup B is drawn in Fig. 5.23. We used the commercial GSM mobile test system TEMS from Ericsson [48] to acquire these additional parameters. TEMS consists of a special GSM mobile station connected via a serial cable to

---

[6] See also Sect. 2.1.

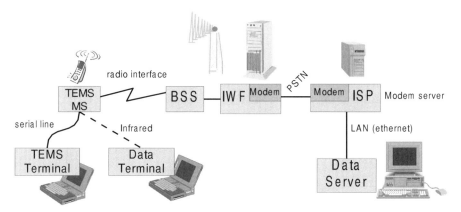

Fig. 5.23. Measurement Setup B for data transmission over GSM CSD traffic channels. The difference from Setup A lies only at the mobile station side

a PC ('TEMS Terminal'). On this PC a special measurement software collects the measurement data incoming from the mobile station on the serial line. The data connection itself is established over the *infrared* connection from the TEMS mobile station to the 'data terminal'. The serial line is *only* used for transfer of measured GSM parameters, and the infrared connection *only* for the actual FTP data transfer. Everything else in Setup B is the same as in Setup A.

With the TEMS mobile station, GSM parameters characterizing the actual radio link state can be analyzed. Among others, the most important ones are RxLev and RxQual. RxLev is the measured signal strength of the carrier, RxQual corresponds to the estimated BER at the mobile station.

5.4.4 Measurement Results for GSM-CSD

The measurements are divided into *stationary* and *wide area* measurements. *Stationary* means, the mobile station was fixed, i.e. $v = 0 \,\mathrm{km/h}$. In *wide area* measurements, the mobile station was mounted in a car while moving around in different environments. We will start with the results of the stationary measurements.

Stationary measurements, FTP transfer. In this subsection we will present results obtained with Setup A. The main results are the net throughput, number of TCP retransmissions and the *overhead rate*, η_o. We define the *overhead rate*,

$$\eta_o := \frac{\mathrm{Throughput[bit/s]}}{\mathrm{PacketLength[bit/Packet]}} \cdot \mathrm{OverheadPerPacket[bits/Packet]} \ , \quad (5.22)$$

as the part of the total throughput that is consumed by the packet overhead. It has the dimensions of a throughput (bit/s). The total throughput thus can be divided into net throughput, which the user observes, and the overhead rate, η_o, which corresponds to the packet header. The throughput divided by the packet length yields the *packet rate*. With this parameter the overhead can be estimated in terms of throughput.

The first results in Fig. 5.24(a) show the mean net data throughput when using FTP data transfer and the SLIP link level protocol. We used TCH/F9.6 in transparent and non-transparent modes. Figure 5.24(b) shows the corresponding TCP retransmissions. Both result plots show the dependence on the TCP packet size. Looking at Fig. 5.24, a maximum throughput of 8.2 kbit/s

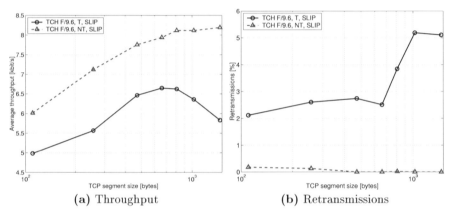

(a) Throughput   (b) Retransmissions

Fig. 5.24. Measured throughput for TCH/F9.6 for transparent and non-transparent mode **(a)**. Figure **(b)** displays the corresponding relative number of TCP retransmissions. Both show the dependence over the TCP packet size

for non-transparent and 6.7 kbit/s for transparent TCH/F9.6 traffic channel can be observed. Thus we can conclude that the non-transparent mode doing local retransmissions with small packet size is much more suited to the highly erroneous mobile radio channel. Because of these local retransmissions, TCP operates on an almost 'error-free' channel. This can be verified by the number of TCP retransmissions shown in Fig. 5.24(b). For non-transparent mode almost no TCP retransmissions are required. Whereas in transparent mode, where all protocol error correction is done by TCP, the number of retransmissions ranges from 2 to 5% of all sent packets.

The dependence on the TCP packet size shows us the expected result. Throughput increases with TCP packet size for non-transparent mode because overhead decreases. As TCP experiences almost no errors, the block length has no influence on the TCP BLER. However, in transparent mode, all ARQ error recovery is done by TCP, and BLER increases with packet length. The maximum throughput has been found at a packet length of $n_D \approx 600$–800 bytes. This measurement result corresponds very well to both theoretical results (Fig. 5.13 for a header length of 40 bytes and $C/I = 20$ dB) and simulated results (Fig. 5.16 $v = 10\,$km/h, $C/I = 20\,$dB). The range of packet length below the optimum one is mainly determined by the protocol overhead. The results beyond the optimum packet length corresponds to high TCP block error rate. This can be verified again by the number of retransmissions in Fig. 5.24(b). Below the optimum packet size the number of retransmissions is about 2% of the total transmitted packets. However, if the packet size increases beyond the optimum one, the number of retransmissions increases significantly to about 5%. In this range the BLER has more influence on throughput which corresponds to the number of retransmissions. Summarizing, the throughput is mainly degraded by the protocol overhead for packet lengths below the optimum one, and by the BLER for packet lengths above the optimum one.

Figure 5.25 shows a comparison between the link layer protocols SLIP and CSLIP. CSLIP is essentially SLIP with active TCP/IP header compres-

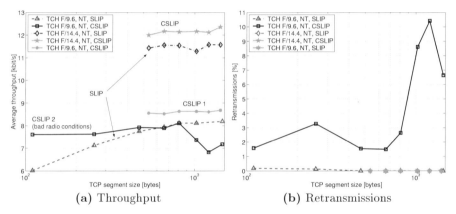

(a) Throughput   (b) Retransmissions

Fig. 5.25. Measured mean throughput **(a)** and relative number of TCP retransmissions **(b)** for SLIP and CSLIP as link layer protocols. Only non-transparent modes for TCH/F9.6 and TCH/F14.4 are depicted

sion. Only non-transparent mode results are plotted, but for both TCH/F9.6 and TCH/F14.4 traffic channels. Again, the left-hand side plot (Fig. 5.25(a)) shows the mean throughput and the right-hand side plot (Fig. 5.25(b)) the relative number of TCP retransmissions. The curve for TCH/F9.6, SLIP is the same as in the previous figure (Fig. 5.24). This illustrates the increasing throughput with increasing packet size. For TCH/F9.6, CSLIP two curves are drawn. One of them (**CSLIP1**) was measured under a normal (good) radio channel. **CSLIP1** yields a mean throughput of about 8.5 to 8.7 kbit/s, which is about 0.5 kbit/s above that for SLIP. Additionally, no degradation for lower packet lengths can be seen, because the influence of header overhead is significantly smaller due to header compression. The second measured curve for TCH F/9.6, CSLIP (**CSLIP2**), experienced poor radio conditions during the measurements. One could clearly see this when looking at the relative number of TCP retransmissions (Fig. 5.25(b)). Curve **CSLIP2** is the only one which has significant retransmissions, ranging from 1.5% to 10.5%. Remember that in Fig. 5.25 only results with non-transparent mode are plotted, i.e. the RLP protocol is active. So even the RLP protocol can not cope with the high BLER alone, and TCP retransmissions are necessary. This corresponds to the observed throughput that is below the one for **CSLIP1** (normal radio channel), but still in the order of TCH/F9.6/SLIP. Nevertheless, we can observe the impact of TCP/IP header compression for very low packet size ($n_D < 500$ bytes). Throughput is substantially higher for **CSLIP2** for $n_D \approx 120$ bytes compared to SLIP, despite the poor radio channel.

The findings for TCH/F14.4 measurements, also drawn in Fig. 5.25, are in principle the same. TCH/F14.4 with SLIP as link layer protocol achieves a mean throughput of about 11.5 kbit/s. TCH/F14.4 with CSLIP yields 12.3 kbit/s. If we compare the relative degradation of both TCH/F9.6 and TCH/F14.4 with respect to their theoretical maximum (without overhead, 9.6 kbit/s and 14.4 kbit/s), the following figures are obtained (Table 5.5):

Table 5.5. Comparison between TCH/F9.6 and TCH/F14.4 in non-transparent mode. Net throughput figures and relative throughput degradation compared to theoretical maximum are listed. Also net throughput plus overhead rate, η_o, and again the relative degradation to theoretical optimum is printed

Traffic channel	Theoretical throughput	Measured throughput and rel. degradation to theoretical throughput (kbit/s)			
		SLIP		CSLIP	
	(kbit/s)	net	net+η_o	net	net+η_o
TCH/F9.6	9.6	8.2	≈ 8.7	8.7	≈ 8.8
		(−15%)	(−10%)	(−9%)	(−9%)
TCH/F14.4	14.4	11.5	≈ 12.1	12.3	≈ 12.5
		(−20%)	(−16%)	(−15%)	(−13%)

Thus TCH/F14.4 has a much higher relative throughput degradation to the theoretical maximum one, -15 to -20%, than TCH/F9.6 (-9 to -15%), when no protocol overhead is considered. This can only be partially explained by the higher overhead obtained because of the higher packet rate, i.e. the 'overhead rate':

The overhead rate, η_o, defined in (5.22), is drawn for all measurements from Fig. 5.25 in Fig. 5.26. When the SLIP protocol was used, the overhead rate consumes roughly 3 to 5 times more bandwidth than with CSLIP. As already pointed out in Sect. 5.3, the overhead decreases with packet size. This behavior is demonstrated in Fig. 5.26. Specifically, the differences between SLIP and CSLIP in throughput in Fig. 5.25(a) are found again in the overhead rate, $\eta_{o,\text{SLIP}} - \eta_{o,\text{CSLIP}}$. The absolute values of η_o and the relative differences between SLIP and CSLIP are listed in Table 5.6. For example, the throughput

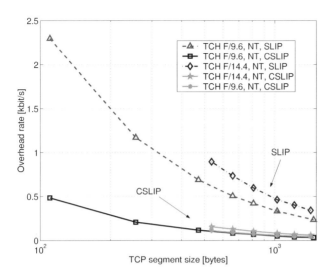

Fig. 5.26. Overhead rate, η_o, for TCH/F9.6 and TCH/F14.4 on the non-transparent mode. SLIP and CSLIP are compared for both traffic channels

Table 5.6. Overhead rate, η_o, for TCH/F9.6 and TCH/F14.4 in non-transparent mode. Both SLIP and CSLIP link layer protocols have been used. This figures are also plotted in Fig. 5.26

Traffic channel	Link layer protocol or difference	Overhead rate, η_o (kbit/s) TCP packet length		
		150	512	1500
TCH/F9.6	SLIP	2.3	0.7	0.3
	CSLIP	0.5	0.2	0.1
	$\eta_{o,\text{SLIP}} - \eta_{o,\text{CSLIP}}$	1.8	0.5	0.2
TCH/F14.4	SLIP	–	0.9	0.35
	CSLIP	–	0.15	0.1
	$\eta_{o,\text{SLIP}} - \eta_{o,\text{CSLIP}}$		0.75	0.25

difference between TCH/F14.4 SLIP and CSLIP is about 0.6–0.8 kbits/s (see Fig. 5.25(a), upper two curves). The corresponding difference in overhead rate, $\eta_{o,\text{SLIP}} - \eta_{o,\text{CSLIP}} \approx 0.25$–0.75 kbit/s (see Fig. 5.26). Thus the difference in throughput between SLIP and CSLIP is almost fully governed by the additional overhead of SLIP. A second example at low TCP packet size: For $n = 150$ bytes the throughput difference SLIP-CSLIP according Fig. 5.25(a) is 1.7 kbit/s. The difference in overhead rate at $n = 150$ bytes is 1.8 kbit/s (see Table 5.6). The agreement is obvious.

Returning to Table 5.5, as already mentioned previously, the overhead does not explain the difference between theoretical maximum throughput and actual observed net throughput. If we add the overhead rate to the net throughput, a relatively large gap to the theoretical throughput still exists. These figures and relative degradations for these cases are also listed in Table 5.5 (columns 'net+η_o'). Thus we conclude that the TCP/IP/SLIP (or CSLIP) protocol overhead is *not* the only bandwidth-consuming mechanism.

Also, the relative degradation is still much higher for TCH/F14.4 (-13% to -16%) than for TCH/F9.6 (-9% to -10%). Thus it also seems that in a good radio channel the theoretical optimum can not be achieved. This is even worse for the TCH/F14.4, because the channel coding for TCH/F14.4 is worse.

Measurement results for High-Speed Circuit-Switched Data (HSCSD) are presented in Fig. 5.27. They have been conducted with the 'Nokia Cardphone 2.0'. Two parallel timeslots are supported, leading to a maximum transmission speed of 28.8 kbit/s with the channel coding scheme of TCH/F14.4. We performed measurements with different link layer protocols, SLIP, CSLIP and PPP. CSLIP and PPP had activated TCP/IP header compression. The

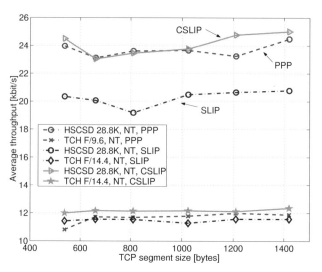

Fig. 5.27. Measurements of HSCSD traffic channel with 28.8 kbit/s theoretical throughput. They are compared with results for TCH/F14.4. Three different link layer protocols have been used, SLIP, CSLIP and PPP

top three curves show the performance of the HSCSD TCH/F28.8 mode. With SLIP, the average net throughput varies from 19.3 to 20.8 kbit/s. Performance with CSLIP and PPP is higher, about 23 to 25 kbit/s. The reason is obviously the activated TCP/IP header compression in CSLIP and PPP. For comparison the results for TCH/F14.4 are also plotted, again with all three link layer protocols.

In both traffic modes (HSCSD TCH/F28.8 and TCH/F14.4), SLIP shows the worst performance, CSLIP and PPP have almost equal performance. However, CSLIP yields almost always a slightly higher throughput. This can be explained by the slightly smaller overhead of CSLIP versus PPP. But this difference is not significant. The relative degradation of HSCSD TCH/F28.8 when using CSLIP/PPP to the theoretical optimum is about 16% ($\approx$ 24.0 kbit/s compared to 28.8 kbit/s). This is almost equal to the degradation observed for TCH/F14.4 and CSLIP when no overhead is considered (see Table 5.5). Therefore this degradation is likely to come from the radio channel quality together with the channel coding performance.

Stationary measurements, RTT evaluation. In this subsection we will present results of *Round Trip Time* (RTT) measurements made with Setup B (see Fig. 5.23). RTT is the time a packet needs to travel from the source to the destination and back again. The measurements have been conducted with the non-transparent TCH/F9.6 traffic channel, i.e. active RLP protocol.

The RTT was measured at the connection between the 'Data Terminal' and the 'Data Server'. The 'Data Terminal' transmits small ICMP packets which are returned by the 'Data Server'. When receiving the response, knowing the sending and receiving time, the RTT can be computed. The Unix program ping was used to perform this task[7]. Simultaneously with the RTT measurements, the 'TEMS Terminal' logged the GSM specific parameters RxLev and RxQual. RxLev and RxQual are defined in [56]. RxLev corresponds to the received signal strength, and RxQual to the estimated BER. RxLev is quantized to 65 values and RxQual to only 8 values. Thus the granularity of RxQual (BER) is rather poor and significance is limited. In all figures we converted RxLev and RxQual already to dBm respectively BER values.

Because of the simultaneous logging of data on two distinct computers ('Data Terminal' and 'TEMS Terminal'), exact clock synchronization was important. This was achieved by the Network Time Protocol (NTP, [99]). NTP synchronizes local clocks to network time servers. But, the timing constraint has been softened because the GSM parameters are averaged over 480 ms (104 TDMA frames) [56] and reported only every 480 ms. Nevertheless, NTP provides relative timing accuracy below 1 ms, which is enough for this measurement.

We executed a large number of measurements to obtain statistically significant results. In Fig. 5.28 we plot the cumulative distribution function (a) and the probability density function (b) of RTT obtained in the measurements. About 90% of the ICMP packets experienced a RTT below 1 s. The mean RTT of these packets is about 900 ms (Cluster 1). This is a very long latency. Previous measurements in [94] reported mean RTTs of about 595 ms. Tracking down the distinct contributions of the RTT, neither the long de-

---

[7] We modified ping slightly to also obtain the time when the ICMP packet was sent. Thus the log files could be synchronized in the off-line processing.

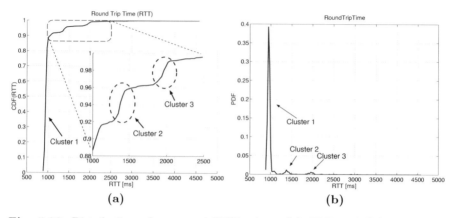

Fig. 5.28. Distribution of measured RTT values, **(a)** CDF and **(b)** PDF. The embedded plot in **(a)** zooms out the interesting details. **TCH/F9.6** traffic channel in non-transparent mode was used

lay nor the difference to the measurements in [94] can be extracted explicitly. However, in [94] the modem server was the actual end of the transmission, *no* additional Internet or Intranet was included. This could be a possible cause for the additional delay of 300 ms in our measurements. This hypothesis is further strengthened by a different measurement report: In [21] mean RTT values of about RTT ≈ 900–1000 ms have also been measured with additional Internet delay.

Theoretically the RTT consists at least of the transmission delay for the ICMP packets (2×77 ms for a 92-byte packet[8]), and Interleaver delay (2×93 ms for TCH/F9.6, [129]) which sums up to a minimum RTT of $RTT_{min} = 340$ ms. The remainder of about 560 ms seems to be introduced in all processing units (base station, IWF, ...), the modem connection between IWF and ISP terminal server and the Intranet of the ISP.

The other 10% of ICMP packets provide deeper insight into the RLP protocol. The embedded plot in Fig. 5.28(a) zooms out the corresponding part of the CDF. Besides the main Cluster 1 at 900 ms, two other clusters exist where an accumulation can be observed. We call them 'Cluster 2' and 'Cluster 3'. Cluster 2 is located around RTTs of about 1400 ms, Cluster 3 around 1800 to 2000 ms. Knowing that the retransmission timeout of the RLP protocol is fixed at 480 ms [64], these clusters can be explained easily. Cluster 2 corresponds to packets suffering from a single RLP retransmission, which adds 480 ms to the base RTT of 900 ms. This gives a total RTT of 1380 ms, which matches measured mean RTT of 1400 ms in Cluster 2 almost exactly. Packets in Cluster 3 suffer from duplicate RLP retransmission, yielding a total addi-

---

[8] The standard ICMP packet used in `ping` consists of 56 bytes dummy payload, 8 bytes ICMP header, 20 bytes IP header and 8-byte PPP framing bytes. This gives a total packet length of 92 bytes.

tional delay of $2 \times 480\,\text{ms} = 960\,\text{ms}$. Adding this to the base RTT of $900\,\text{ms}$ we obtain $1860\,\text{ms}$. Again, this is in the range of Cluster 3.

Thus we conclude that approximately 90% of all RLP packets are transmitted without errors, 4–5% experience a single retransmission at the RLP layer and 3–4% suffer from duplicate RLP retransmissions. The remainder of 1–2% are delayed by 3 or more retransmissions.

In Fig. 5.29 we plot the evaluation of RTT measurements for the HSCSD traffic channel TCH/F28.8. Again CDF and PDF distribution functions are

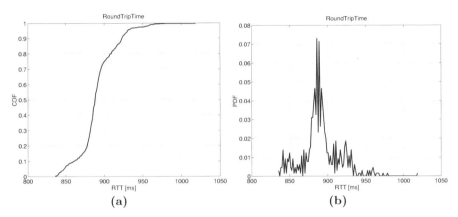

Fig. 5.29. Distribution of measured RTT values, **(a)** CDF and **(b)** PDF for the **HSCSD** traffic channel **TCH/F28.8**

plotted. The majority of all measurements are around RTT=$880\,\text{ms}$. This is only some $20\,\text{ms}$ below the value for TCH/F9.6. However, quite a number of measurements (about 10%) have been obtained with an RTT below $\approx 880\,\text{ms}$. This was not the case in TCH/F9.6. Also at higher RTTs a much more smoothed distribution is obtained for RTT>$900\,\text{ms}$. No additional explicit clusters, besides the main cluster mentioned before, can be observed for higher RTTs. No RLC retransmissions due to timeouts occur at all, because these would increase the RTT by at least $520\,\text{ms}$[9]. Such high RTT ($880\,\text{ms}+520\,\text{ms}$) have not been measured.

It seems that the smoothing of the RTT comes from the external network (modem link or internal ISP network).

In the following we will present results of combining the GSM parameters RxLev and RxQual to the RTT. The combining yields (RxLev,RTT) and (RxQual,RTT) combinations. A pair is determined by the same absolute time when both values have been measured. A plot of all (RxLev,RTT) combinations is drawn in Fig. 5.30(a). Again all three clusters of RTTs are visible,

---

[9] In the 14.4-kbit/s channel coding scheme the RLP retransmission timeout has a default value of $520\,\text{ms}$ rather than $480\,\text{ms}$ for TCH/F9.6.

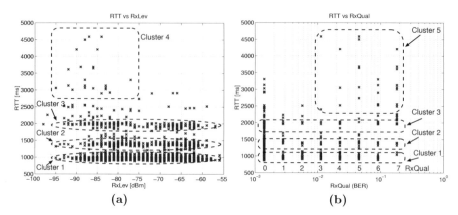

Fig. 5.30. Pairs of (RTT,RxLev), **(a)**, and (RTT,RxQual), **(b)**, combinations with equal absolute time

now distributed over the range of measured signal strength levels (RxLev). No significant concentration of RTTs on specific RxLev values can be found. The distribution of all three clusters is rather homogeneous over the RxLev range. This is not very surprising, because usually a GSM system is an interference limited system. Thus signal strength is not *the* significant measure but rather BER (RxQual), as long as signal strength is above a minimum reception level. Signal-to-noise and interference ratio, SNIR, mainly determines the reception quality, where interference has precedence. However, very large RTTs, RTT > 3 s only occur at RxLev < 82 dB (Cluster 4). This means the probability of very high RTTs correlates with RxLev. This result was somewhat unexpected, also due to the relatively high RxLev values. It indicates that the probability of low transmission quality grows at low signal strength, *despite* the system being interference limited. But at low signal strength the system seems to change at least partly to a noise-limited one.

Combining RxQual values with the corresponding RTT values results in Fig. 5.30(b). Most of the RxQual values are at BER $\approx 10^{-3}$ (RxQual=0). The three clusters of RTTs (corresponding to base RTT, RTT with single and double RLP retransmissions) can be identified again. Although at the lowest BER measurement value (BER $\approx 10^{-3}$) this is not visible at first sight. But another cluster can be observed (Cluster 5). It surrounds (RxQual,RTT) combinations with high RTT and also high BER. That is, the probability for very high RTTs increases with high BER. This observation corresponds to the expectation that poor BER gives more often RLP retransmissions and subsequently higher RTT. Surprisingly, this behavior starts rather late only for higher RTTs (RTT > 2–3 s), which are already rather long RTTs.

Summarizing Fig. 5.30, the best performance is achieved at a BER < 2×10^{-2} (RxQual<4). If quality is worse (RxQual$\geq$4) then data transmission performance is reduced. This results in spuriously large RTTs.

Stationary measurements, Sample sequence trace. To provide further insight into the behavior of TCP and RLP in non-transparent mode, we will present an analyzed data transmission sample. We performed a downlink data transmission from the 'Data Server' to the 'Data Terminal'. Basically Setup A was used, but an additional measurement tool at the A-interface (see Fig. 3.1) was used. Thus two simultaneous measurements at the TCP layer and at the RLP layer were performed.

The program `tcpdump` was responsible for logging the TCP packets, as in all previous measurements. `tcpdump` was running at the 'Data Terminal' on the mobile station side of the connection. The RLP packets have been logged with the special measurement tool at the A-interface between BSC and MSC.

Because multiple layers are traced simultaneously this measurement could be called 'multi-layer tracing', similar to the analysis in [95]. The results from [95] have been already presented in Sect. 5.1.3.

The evaluation of both data logs are plotted in Fig. 5.31.

Only the end of the entire data transmission is shown, i.e. transmission of the last 15 TCP packets (excluding retransmissions) corresponding to 22 500 bytes (at 1500 bytes packet size). On top (Fig. 5.31(a)) the TCP trace is drawn. It shows the sequence number in bytes over time. All relevant details are braced by the TCP sliding window, which determines the lower and upper bound of valid TCP segment numbers. The sliding window's upper end is determined by the advertised window, which is illustrated by the light dotted line in the upper part of Fig. 5.31(a).

In the bottom, the sliding window is bounded by the sequence number of the highest correctly and in-sequence received TCP packet. A received TCP packet is represented by a '$\updownarrow$' sign, or by an additional '$\Diamond$' sign on top of the double arrow, if the TCP push flag is set (see legend in Fig. 5.31(a))[10]. Every time a correct in-sequence TCP packet is received, this is raised by one packet size (compare with area around $t = 180$ s in Fig. 5.31(a)). For $t < 181$ s, every packet was received correctly, no error occurred. Slightly later, a TCP packet is missing; it has not been received at the mobile station terminal. Thus, all TCP packets arriving later do trigger an acknowledgment packet (as usual), but the sequence number of these acknowledgment packets does not change, i.e. the lower end of the sliding window does not change. This is indicated by the small tick marks around the area $t \approx 190$ s. The number '3' at the third acknowledgment packet with the same sequence number illustrates the third duplicate acknowledgment which triggers the fast retransmission algorithm at the TCP sender.

Subsequently this continues until the last TCP packet is received, i.e. the 'FIN' bit is set. This tells the receiver that the last packet of the actual data sequence has been received.

---

[10] The TCP push flag tells the receiver that it should handle the received data immediately up to the application. It has no influence on the protocol mechanisms.

sequence number

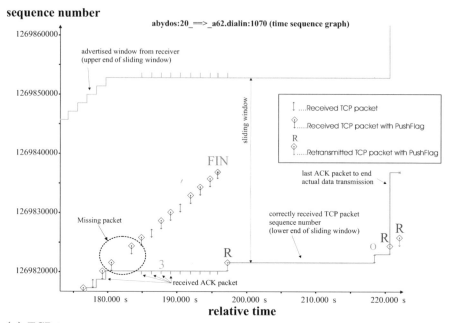

(a) TCP trace

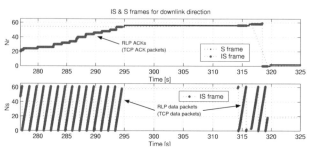

(b) RLP trace, downlink direction

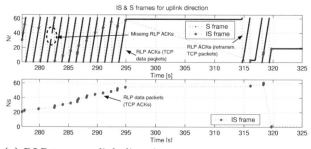

(c) RLP trace, uplink direction

Fig. 5.31. Analyzed TCP **(a)** and RLP traces for downlink **(b)** and uplink **(c)** direction. The TCP trace shows the TCP sequence numbers over the time. In the RLP traces the sending RLP sequence number N_s and the receiving (acknowledge) sequence numbers N_r are plotted for uplink and downlink direction

However, because 'fast retransmission' was activated at the TCP sender, retransmitted packets follow after the 'FIN' packet. Only when this packet is received is the acknowledgment number increased and sent back to the sender. But it took too long a time until the ack number was raised, so a timeout occurred at the sender and the missing packet was transmitted rather late at $t \approx 218$ s. So there was a gap in transmission of approximately 20 s where no data was transmitted. Also, the next two packets are retransmitted although they have been received correctly at their first transmissions. This is because the sender does not know, due to the outstanding acknowledgment, which had not been yet received (go-back-N mechanism).

Analyzing the RLP frames we find the cause of the TCP packet loss. In Fig. 5.31(b) we plot the *downlink* transmission sequence number N_s and the acknowledging (receiving) sequence number N_r over the time. Note that the absolute time was *not* synchronized in this measurement between TCP and RLP logging. The same for the RLP frames in the *uplink* direction is shown in Fig. 5.31(c).

In the downlink direction, N_s shows the sequence of transmitting the TCP data packets, in uplink the N_s sequence corresponds to the TCP acknowledgments. Only RLP frames carrying information (I+S frames) have a N_s sequence number. I+S frames include also an acknowledging sequence number N_r (combined information and supervisory frame). Special supervisory frames (S-frames) are used for acknowledging RLP data frames (I+S) and include only a N_r acknowledging number. That is, the S-frames in uplink acknowledge the I+S-frames in downlink which carry the TCP data packets (compare N_r of S-frames in Fig. 5.31(c) with N_s of I+S-frames in Fig. 5.31(b)). The same is true vice versa for TCP acknowledgment packets, S-frames in downlink acknowledge I+S-frames in uplink.

However, if an I+S frame is sent anyway, because there is data to send available, *no* explicit S-frame is transmitted. Rather the N_r field in the I+S-frame is used for acknowledging. This is indicated by the large marks in the N_r plots (corresponding to I+S-frames) and small marks (corresponding to S frames). Summarizing, for acknowledging (N_r) I+S RLP frames either I+S-frames in the opposite direction (piggybacked at an information frame) or only S-frames are used.

Looking at the N_s plot in Fig. 5.31(b) (downlink) we find a large gap in data transmission between $t \approx 295$ s and $t \approx 314$ s. This corresponds to the gap in Fig. 5.31(a) (TCP trace) between $t \approx 197$ s and $t \approx 218$ s where no TCP data packets are transmitted at all.

Taking a closer look at the acknowledgment RLP frames (N_r in Fig. 5.31(c)), we find RLP acknowledgment packets missing. As the RLP frames have been logged at the A-interface, this means that either RLP data packets (downlink, MSC to MS) or RLP acknowledgment packets (uplink, MS to MSC) have been lost at the radio interface. A zoomed plot shows this part in more detail (Fig. 5.32). These missing RLP acknowledgments

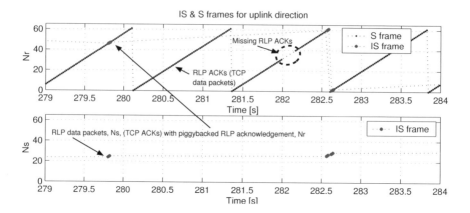

Fig. 5.32. Zoomed part of uplink RLP frame trace from Fig. 5.31(c)

are the cause of the missing TCP packet in Fig. 5.31(a), which leads to the retransmissions of TCP packets and the timeout at the TCP sender. A total of 8 RLP frames are not acknowledged, i.e. $8 \times 192\,\text{bits}=192\,\text{bytes}$. This is only $\approx 13\%$ of a TCP packet but nevertheless it drops the entire TCP packet.

Normally RLP retransmissions should recover such errors. But in this case, no recovery occurred. We were not able to determine the cause for this.

Wide area measurements. We performed *wide area* measurements for the conventional TCH/F14.4, the TCH/F19.2 and TCH/F28.8 HSCSD traffic channels[11]. A few measurement results for TCH/F9.6 are already available ([21, 77, 88]), thus we concentrated on the other traffic channels. Non-transparent mode was set, i.e. the RLP protocol tried to provide a reliable channel to higher layers (TCP/IP). The data link layer protocol was PPP with activated header compression. We used measurement Setup A. The laptop was mounted in a car. Nokia's cardphone was connected to the laptop. The measurements were conducted in the city and the surroundings of Vienna, Austria. We evaluated the TCP net throughput during an FTP connection and the number of retransmissions of TCP packets.

To assess TCP throughput performance in different environments, we selected five scenarios [77]:

- Rural area, parking place: This scenario was selected to get a noise-limited situation. Mobile station velocity was zero.
- Rural area, standard road: Mobile station velocity varied between 50 km/h and 100 km/h. Terrain was flat, no hills or mountains are located in this area. Cell size is large, thus only infrequent handovers occurred.

---

[11] The TCH/F19.2 traffic channel occupies 2 timeslots and uses the same channel coding as TCH/F9.6.

- Urban area: This scenario is characterized by a relatively small cell size and a high number of users. Mobile station velocity varied between 0 km/h and 50 km/h, car traffic dependent. Link level performance is usually interference-limited.
- Highway: This scenario includes high mobile station velocities from 100 km/h up to 130 km/h. The highway is located in a rural area, thus the cell size is large.
- City highway: This highway is located directly in the city. Mobile station velocity was about 80 km/h. As cell size is small in the city, handovers occurred more frequently.

In every environment at least 10 different measurements were performed. However, in two environments ('City Highway' with TCH/F14.4 and 'Highway' with TCH/F19.2) *no* measurement results were obtained. In these cases no full FTP data transfer was possible due to poor link quality.

General comments:
In general, it was hardly possible to set up a connection *during movement*. This problem was particular severe for TCH/F14.4 and TCH/F28.8. Both traffic channels use not so powerful channel coding.

The main problem was, in 98% of the cases, *not* the establishment of the GSM link. GSM link establishment worked very reliably, also in poor environments and with worse channel coding schemes (TCH/F14.4 and TCH/F28.8). Also *no* differences in blocking probability with HSCSD links (TCH/F19.2 and TCH/F28.8) compared to non-HSCSD links (TCH/F14.4) have been observed. Thus *no* capacity problems seem to exist in this particular network.

Instead, the *real* problem was to log on to the fixed network server, verifying user-name and password and setting up a PPP connection. For authentication the *Password Authentication Protocol* (PAP) was used. After successful authentication, PPP performs negotiation of some protocol parameters. We were not able to track down in which particular phase of connection establishment most of the errors occurred. However, it seems that either one or both of these two protocols are able to cope only in a limited way with frequent timeouts, unreliable connections and short disconnections due to handovers.

Results:
The performance of TCH/F14.4 in the different scenarios is shown in Fig. 5.33. The left-hand side plot shows minimum, mean and maximum observed throughput, the right-hand side plot the relative number of TCP retransmissions. For the 'city highway' scenario, *no* results were obtained. The observed mean throughput does not differ significantly between the environments. It is rather constant at about 10 kbit/s. Comparing this with the results of the stationary measurements (Fig. 5.25) a degradation of almost 20% is found. Variation in throughput is largest for the 'urban area' scenario, minimum throughput variation is obtained for 'rural area, parking place'. But only in the 'urban area' scenario did we measure throughput values near that mea-

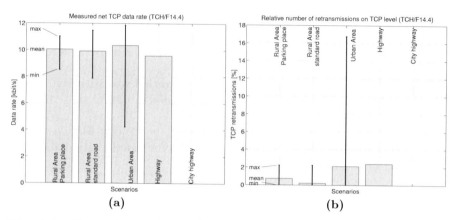

Fig. 5.33. Measured throughput (**a**) and relative number of TCP retransmissions (**b**) of **TCH/F14.4** in different environments

sured in the stationary measurements (Fig. 5.25). In the 'Highway' scenario too few measurements were available, thus no variation is plotted.

Looking at the TCP retransmissions, we find no direct correlation to the throughput mean values. But TCP retransmissions are very low for the two rural area scenarios and significantly higher in the 'urban area' and 'highway' scenarios.

Results for TCH/F19.2 are drawn in Fig. 5.34. For this traffic channel no measurement results for the 'Highway' scenario are available. Again the mean throughput shows not much difference between all scenarios. It varies from 14 kbit/s to slightly more than 15 kbit/s. This corresponds to 72% (78%) of the theoretical maximum. But variation is largest for 'Rural Area, standard

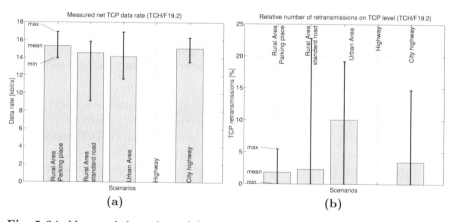

Fig. 5.34. Measured throughput (**a**) and relative number of TCP retransmissions (**b**) of **TCH/F19.2** in different environments

road' and 'Urban area'. It is interesting that the variation in the 'City high-way' scenario is almost equal to that of 'Rural area, parking place', and below the other ones.

The number of TCP retransmissions is, in general, higher than for TCH/F14.4 and also 'Urban area' and 'City highway' have the largest mean number of TCP retransmissions. However, now the variation in the 'Rural area, standard road' scenario is also very large.

Finally, measurement results for TCH/F28.8 are presented in Fig. 5.35. One special feature of the TCH/F28.8 settings was the automatic fall-back to

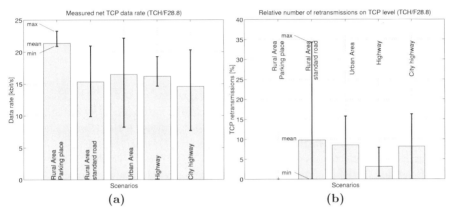

Fig. 5.35. Measured throughput **(a)** and relative number of TCP retransmissions **(b)** of **TCH/F28.8** in different environments

TCH/F19.2 when link quality falls below a specific threshold. Thus the results for TCH/F28.8 reflect not only the quality of TCH/F28.8 alone but the joint performance of TCH/F28.8 and TCH/F19.2. This has to be considered when we assess the performance of TCH/F28.8.

The mean throughput for the 'Rural area, parking place' scenario is sig-nificantly higher than for all other scenarios. It is more than 21 kbit/s. Mean throughput in all other scenarios is of the same order of magnitude as for TCH/F19.2, but slightly higher, about 15 kbit/s to 16 kbit/s. Comparing these results again to the stationary measurements (Fig. 5.27) the through-put is from about 13% (for 'Rural area, parking place') up to 37% (for all other scenarios) lower in the wide area measurement scenarios.

Throughput variation is, in general, very high except for scenario 'Ru-ral area, parking place'. This, and the almost equal mean throughput on 4 scenarios indicate frequent 'fall-backs' to TCH/F19.2 during TCH/F28.8 measurements.

The number of TCP retransmissions is again significantly higher than for TCH/F19.2 and ranges from 3% up to 10%. But there is one exception, scenario 'Rural area, parking place' shows *no* TCP retransmissions at all.

This indicates very good link quality, but still not as good as in the stationary measurements in the urban area (section 'stationary measurements').

To summarize:
Mean throughput performance in more challenging scenarios is from 13% up to 37% *lower* than in stationary, urban area environment. In general *no* significant difference in performance between different scenarios was observed, with one exception: The stationary measurements on a parking place in the rural area performed significantly better. The number of TCP retransmissions was, in general, quite high. Recall that GSM non-transparent mode was used. This indicates that the GSM RLP protocol was in most of the cases *not* able to provide a reliable link. Instead TCP has to take care and retransmits packets in error.

5.4.5 Summary and Conclusions

We presented throughput and delay investigations of the GSM CSD traffic channels by means of both simulations and measurements. We considered a typical FTP file download. A user wants to download a file from a server in the Internet via the GSM radio interface.

In simulations we used TCH/F9.6 and TCH/F14.4 traffic channels and SLIP as link layer protocol. Only GSM transparent mode simulations were carried out. In the measurements TCH/F9.6, TCH/F14.4 and TCH/28.8 have been investigated. Both transparent and non-transparent modes have been used. As link level protocols we used SLIP as well as PPP.

TCP throughput and delay. In a first step we investigated the end-to-end throughput and delay performance of the TCP protocol.

As the GSM measurement tool that we used (TEMS, [48]) does not support C/I measurements, simulations are very important to assess the performance for various C/I values. In the measurements, C/I can neither be set explicitly nor measured.

An important result that we obtained from the simulations can be drawn from Fig. 5.15. This shows the *minimum* C/I necessary for acceptable throughput $(C/I_{\min})$. Remember that this figure is valid for transparent mode only, which explains the relatively high C/I values. $C/I_{\min}$ depends heavily on the traffic channel and the mobile station velocity. The threshold for TCH/F9.6 is about $C/I_{\min} \approx 15\,\mathrm{dB}$ to obtain about 67% of the maximum throughput ($8\,\mathrm{kbit/s}$). For TCH/F14.4 roughly $5\,\mathrm{dB}$ more in C/I is necessary to obtain about 80% of the maximum throughput ($\approx 13\,\mathrm{kbit/s}$).

Going to a mobile station velocity of $v = 150\,\mathrm{km/h}$, TCH/F9.6 needs almost the same $C/I_{\min} \approx 15\,\mathrm{dB}$ as for $v = 10\,\mathrm{km/h}$ to obtain 67% of maximum throughput. Also maximum throughput has not decreased significantly. But the slope of the curve is steeper, thus throughput will be more affected by deep fades or shadowing effects. TCH/F14.4 loses more peak throughput but threshold $C/I_{\min}$ remains almost constant.

We conclude:
Transparent TCH/F9.6 is suitable for data transmission up to high mobile station velocities, and for poor radio environments ($C/I <$ 15 dB). However, if C/I is above about 15 dB, TCH/F14.4 *should* be preferred. Throughput is higher for TCH/F14.4 than for TCH/F9.6 in almost all cases, despite the fact that TCH/F14.4 can *not* be fully utilized to its optimum value in general.

From a users point of view, if costs/min are equal for TCH/F9.6 and TCH/F14.4, TCH/F14.4 should be used. From an operators point of view, TCH/F14.4 can be treated equally to TCH/F9.6, as for both only a single timeslot is used. No capacity is lost, but perceived quality of service by the end user is higher. Thus the introduction of TCH/F14.4 has an advantage for operators and end users.

The TCP end-to-end delay presented in this section is measured for a bulk data transfer (e.g. FTP file download). Because TCP fills up the buffers located in the transmission path up to a stable value, the end-to-end delays are rather high. We experienced exactly this behavior in Figs. 5.17 and 5.18, where the IWF buffer size corresponds to the delay. This delay will not occur for interactive traffic, where only a few packets are traveling from transmitter to receiver. The buffers can not be filled up with these few packets, therefore the delay observed by the user will be of the order of the RTT. This was according to the measurements about RTT $\approx$ 900 ms.

The mean throughputs for TCH/F14.4 and TCH/F28.8 have been measured. Compared to the theoretical throughput, both traffic channels show more degradation than TCH/F9.6. The degradation is of the order of 15–20% for TCH/F14.4 and TCH/F28.8, whereas for TCH/F9.6 only a relative degradation of 9–15% was observed.

This is an indication that the worse channel coding can *not* cope with the radio channel as well as the channel coding for TCH/F9.6.

Optimum TCP packet length. In transparent mode the mean throughput depends more heavily on the TCP packet size than in non-transparent mode. We investigated this topic theoretically in Sect. 5.3. For small packet size the overhead reduces the throughput. For large packet size, an error on the channel effects a larger block, which needs to be retransmitted. Again this decreases throughput. Therefore an optimum packet length exists.

The simulation results agreed with the theoretical computation very well (compare Fig. 5.16 with Fig. 5.13, for a header length of 40 bytes). Also the measurements agreed with the theoretical and simulation results (compare also Fig. 5.24).

In non-transparent mode, the RLP protocol provides a virtually error-free channel to higher layers. Thus almost no retransmissions on TCP level occur. Therefore, the total throughput is maximized, when the overhead is minimized, i.e. for large packet sizes.

But as the measurements showed the RLP protocol is in many cases not able also to provide an error-free link. Thus, on a poor channel, the TCP packet length might be an important parameter also with the non-transparent mode (active RLP).

We conclude:

The optimum TCP packet length for *transparent* TCH/F9.6 traffic channel in GSM is, on average, around 700 bytes. When the channel is worse the optimum packet length shifts to lower values (down to ≈ 500 bytes). For better channel conditions a larger block size is optimal (up to ≈ 900 bytes).

In *non-transparent* mode, the TCP packet length should be as large as possible, as long as the RLP protocol can cope with the errors. Otherwise, TCP packet length should also be of the order of the figures suggested for transparent mode.

Link layer protocol influence. We performed measurements with SLIP (no TCP/IP header compression), CSLIP (activated header compression) and PPP (activated header compression) as link layer protocols. TCH/9.6, TCH/F14.4 and also HSCSD TCH/F28.8 have been measured. To determine the overhead in comparison with the throughput, we defined an *overhead rate* (5.22). This corresponds to the fraction of the total throughput consumed by the overhead of the packets.

The overhead rate of CSLIP is smallest compared to PPP and especially SLIP, thus the throughput in non-transparent mode is highest. For very low TCP packet size, header compression increases throughput significantly. Thus the overhead is also much smaller for small packet sizes (compare Fig. 5.25 and Fig. 5.26). Small packet size can occur frequently in Web browsing applications and in the reverse path (acknowledgment packets). Page requests have typically very small size starting from some 50 bytes.

The overhead rate also grows when throughput is increased, because more packets are transmitted.

The PPP protocol has TCP/IP header compression included and showed almost equal performance to CSLIP. However, at TCH/F14.4 the performance of PPP was slightly below that of CSLIP (see Fig. 5.27).

We conclude:

Header compression should be used in any case, at least in non-transparent mode. CSLIP and PPP are almost equal in throughput performance, although CSLIP might perform better in some cases due to slightly less overhead. However, PPP provides more sophisticated link establishment, which is more convenient to use for the user. Thus PPP might be the better choice.

RTT evaluation. We evaluated a large number of RTT measurements for both TCH/F9.6 and TCH/F28.8. Both traffic channels operated in non-transparent mode. 90% of the samples for TCH/F9.6 showed a RTT of

RTT $\approx 900\,\text{ms}$. This is roughly $300\,\text{ms}$ more than reported in previous measurements [94], but there without external Inter- or Intranet. However, in a different measurement study [21], these high mean RTT values (around $900\,\text{ms}$) have also been obtained. This is a rather high RTT which disturbs especially highly interactive traffic, like `telnet` sessions.

From the distribution of the RTT samples we extracted roughly the number of RLP retransmissions which occurred due to timeouts. About 90% of all RLP packets experienced no retransmission, 4–5% a single retransmission and 3–4% suffered from duplicate RLP retransmission. The remainder (1–2%) of the packets experienced more than three retransmissions.

A somewhat different result was obtained by RTT measurements of TCH/F28.8. There the mean RTT of more than 50% of the ICMP packets is about $880\,\text{ms}$. This is only slightly below the mean RTT observed for TCH/F9.6. Thus the radio interface itself seems to have only a minor influence on the delay of the packets. The difference is exactly the difference in transmission speed. Also *no* RTT values higher than $1000\,\text{ms}$ occurred. This lets us conclude that *no* retransmissions due to timeout occurred at all.

We combined the RTT samples with simultaneously measured GSM radio link parameters, RxLev, corresponding to the field strength and RxQual which corresponds to the estimated BER. The clusters observed in the RTT sample distribution can be found again in Fig. 5.30. However, no correlation between RxLev and RTT can be observed with one exception: Very high RTTs (RTT>3 s) occured *only* at field strength levels below -75 to $-80\,\text{dBm}$. Thus, in noise-limited environments, the probability for very high RTTs (corresponding to more than 3 to 4 RLP retransmissions) is much higher than in interference-limited environments.

Also, in general *no* significant correlation between RxQual and RTT can be measured. Again there is one exception for very high RTTs: RTT samples above RTT>2.5 s occured *only* for bit error ratios higher than BER $> 2 \times 10^{-2}$ (RxQual$\geq$4). Thus for high BER the probability for high RTT is increased significantly.

Sample sequence trace. We showed, by tracing TCP and RLP packets simultaneously, the interaction between these two protocols. Errors at the radio interface which are not recovered by RLP cause errors and trigger recovery procedures at the TCP layer. The mutual dependency has been shown in detail (Fig. 5.31).

A source of long transmission gaps are the TCP timeouts, which tend to be very long because of TCP's own RTO computation method (RTO $=$ $\overline{\text{RTT}} + 4\sigma_{\text{RTT}}$). A possible solution would be to *adapt* the retransmission timeout calculation to the environment.

Wide area measurements. We conducted wide area measurements with non-transparent GSM mode for TCH/F14.4, and the HSCSD traffic channels TCH/F19.2 and TCH/F28.8. We investigated five different scenarios: 'rural area, parking place', 'rural area, standard road', 'urban area', 'highway' and 'city highway'.

In four of the five scenarios the mean throughput was about 20% to 37% *lower* than throughput observed in the stationary measurements in urban area. This corresponds to the relatively high mean number of TCP retransmissions observed up to 10%. The only exception is the 'rural area, parking place' scenario, where only a degradation of 13% compared to stationary measurements together with *no* TCP retransmissions was observed.

Thus the RLP protocol is *not* able to provide a reliable link in most of the cases. A higher-layer protocol (e.g. TCP) needs to provide reliability as well as to ensure total correctness of the transmitted data stream.

The mean number of TCP retransmissions is significantly higher for HSCSD traffic channels (up to 10%) as for the non-HSCSD TCH/F14.4 traffic channel. Note that TCH/F28.8 and TCH/F14.4 utilize the same channel coding scheme. It seems that the multi-link version of RLP (Version 2) does *not* have the same error correction capability than RLP for single links (Version 0 and 1). However, without deeper investigations of the RLP protocol behavior no reliable statement is possible.

A large problem during the measurements was the entire connection establishment. However, GSM link establishment worked very reliably; rather the network login and the PPP connection setup were the problem areas. Either the *Password Authentication Protocol* (PAP) used for checking authentication and log on to the server or PPP during its parameter negotiation phase has serious problems with frequent disconnections and long timeouts.

We conclude:

Throughput performance on more challenging scenarios is up to 37% below that in stationary, urban area scenario. The RLP protocol for HSCSD seems to be not as powerful as for the single link traffic channels and needs further investigation.

HSCSD TCH/F28.8 traffic channel has almost always higher throughput than TCH/F19.2 despite the frequent automatic fallback into the TCH/F19.2 traffic channel. Thus we recommend use of TCH/F28.8 when available, despite the worse channel coding.

Higher layer protocols needed for network log on, like PAP, and the link layer protocol PPP during its connection setup phase introduce serious problems during connection setup. Further investigation on these protocols concerning robustness against frequent disconnections and long TCP timeouts are necessary to retrieve the real cause of the connection setup problem.

5.5 TCP over GSM-GPRS, Simulations

In this section we will present simulations of TCP/IP performance over GSM GPRS. As for GSM-CSD, we performed all simulations in 'Ptolemy' ([78]).

In the beginning we will present the simulation model. The simulation results will follow in the subsequent section.

5.5.1 Simulation Model for GSM-GPRS

The simulation of GPRS is more challenging than that of GSM-CSD. GPRS is completely packet switched, also over the radio interface. Multiple users share the same radio resources. A completely new dimension must be considered for simulation of multiple users.

In general, our simulations can be divided into *single-user* and *multiple-user* models. When modeling only a single user, the radio resources are exclusively reserved for this user. Multiple users have to share the available radio resources among each other. The radio resources are in essence the k timeslots of a carrier, which carry a *Packet Data Channel* (PDCH). The number of PDCHs on a carrier is definable by a parameter, but is fixed during a simulation. We assume an already established connection, i.e. the mobile station is in 'ready mode' where data can be sent and received immediately. No mobility management is taken into account [63]. Only a single carrier frequency with 8 timeslots per frame is considered. Considering more than one carrier in the simulation would *not* bring additional insights but only a scaling of the results presented in the following, *if* the number of control channel per carrier remains the same. At the end we will also present a model for simulation of a combined circuit-switched, packet-switched network, where CSD and GPRS users share the same resources. There also we simulate only a single carrier because of the huge simulation effort, both in computation time and storage space.

The detailed simulation model for the **single user** case is shown in Fig. 5.36. As in GSM-CSD we assume a standard scenario: downloading data from a fixed server in the Internet to a GPRS mobile station. In the *single user model*, we assume a data source sending at *constant rate*. Thus there is always enough data available for sending at the TCP sender. This we can think of as FTP or bulk data transfer. If only a single user is demanding radio resources, we can determine the maximum possible throughput a user experiences under 'ideal' conditions. 'Ideal' in this context means that no collisions or bandwidth sharing with other users occurs.

The TCP sender transmits the data packets and sends them one layer down to IP. We assume further that we can model the complete Internet and backbone network of the mobile network by a single transmission line. Thus the GGSN may be omitted as independent node in the model. The line has a bandwidth of $B = 2\,\text{Mbit/s}$ and a delay of $\tau = 10\,\text{ms}$. IP does not fragment

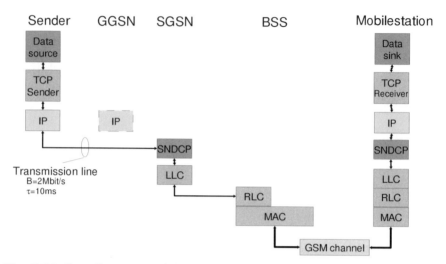

Fig. 5.36. Overall structure of the model for simulation of TCP/IP data transmission over GPRS assuming a single user

the TCP packets. The maximum IP packet size is set to 65 535 bytes. TCP packet length is always below this limit [111].

The IP packets arrive subsequently at the SGSN and are segmented into smaller packets according to the maximum LLC layer packet size. This is set to 1500 bytes. No TCP/IP header compression and no data compression is activated in the SNDCP layer.

The LLC layer frames the SNDCP packets with a header. LLC operates in *unacknowledged mode*, so no additional retransmissions are introduced at the LLC layer. However, the LLC check sum is computed and inserted in the LLC frame. It is checked at the receiver. If a corrupt packet is detected (wrong CRC), then the packet is discarded. The LLC packets are handed down to the RLC layer.

RLC can operate in both acknowledged and unacknowledged mode. It fragments the incoming LLC packets into smaller RLC packets. RLC packet size depends on the coding scheme. We implemented all four coding schemes, CS-1 to CS-4. At the RLC receiver the RLC packets are concatenated to LLC packets again. RLC is a selective repeat ARQ protocol with a window size of 64. Thus, in acknowledged mode, an inner loop exists between the two RLC entities at base station and mobile station side.

The MAC layer multiplexes the RLC packets on the available PDCHs. For multiplexing the multislot capabilities of the simulated handset have to be taken into account. All combinations of uplink and downlink multislot capabilities can be set. We will focus in the simulations on the multislot capabilities in the downlink direction (base station transmitter – mobile station receiver), because we consider only downlink data transfer. We denote the

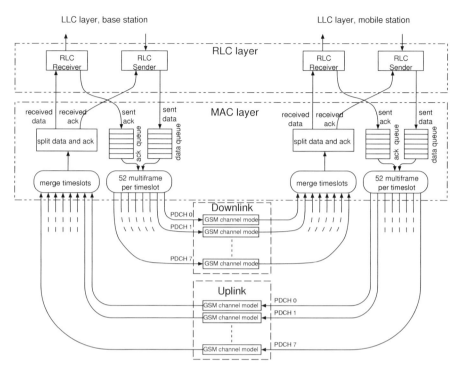

Fig. 5.37. Detailed RLC and MAC layer of simulation model

mobile station receiver multislot capability with 'RxTS'. A detailed illustration of the RLC and MAC layer for a single user is shown in Fig. 5.37. The data flow from base station to mobile station (downlink path) is as follows: The LLC packets from the base station LLC layer arrive at the RLC sender at the base station (upper left-hand side RLC sender in Fig. 5.37). The RLC sender segments them into smaller packets and sends them to the *data queue* of the MAC scheduler. Parallel to the data queue an *acknowledgment queue* is maintained. The RLC receiver at the base station has to acknowledge the data packets received from the uplink. These acknowledgment packets are sent on the downlink path.

The MAC layer has the responsibility of scheduling the RLC/MAC packets onto the available timeslots. A timeslot corresponds to a packet data channel, PDCH. We implemented a simple queuing mechanism, where the acknowledgment queue has priority. There, RLC data packets are only sent, when *no* acknowledgment packets are waiting in the acknowledgment queue. If a PDCH is available, the oldest packet in the queues (acknowledgment before data) is sent according to the 52-multiframe for the actual PDCH. The number of available PDCHs is an input parameter and is fixed during a simulation run. Every PDCH (i.e. each timeslot) is transmitted over a separate channel model. We used the GSM channel model introduced in Sect. 3.4. Subsequently, when the RLC packets arrive at the mobile station, all packets of all PDCHs are merged together. Then the packet stream is split into data and acknowledgment packets. Finally, data packets are handed to the RLC receiver and the acknowledgment packets to the RLC sender.

The RLC packet flow in uplink direction is exactly the same.

When multiplexing the packets on the 52-multiframe, a *Packet Random Access CHannel*, PRACH, is modeled. We assume one packet random access on one PDCH out of all PDCHs for one multiframe period. That is, if there exist 8 PDCHs, a random access is modeled only on the first PDCH once for every multiframe period. A packet random access lasts 4 GSM frames (equal to one RLC frame).

In the uplink, no dynamic resource allocation is done. We assume there is always enough resources available in the uplink. Nevertheless, this does not influence the simulation results seriously, as we assume only downlink data transfer. Since only TCP acknowledgment packets are transmitted in the uplink, no congestion will occur at the uplink.

For modeling **multiple users** sharing the same PDCHs, nothing is changed in principle. Only the user specific parts are duplicated for every user. These are the data source, TCP, IP, SNDCP, LLC and RLC layers. However, MAC layer and the channel are the same for all users. See Fig. 5.38 for an illustration. Every user has a completely separate data path. Mobile station 1 has a connection to Server 1, MS 2 to Server 2, and so forth. There is no data exchange between the different user paths. Each protocol stack

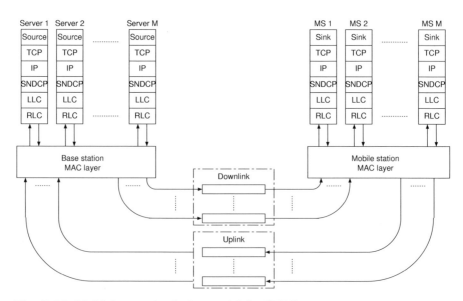

Fig. 5.38. Multiple user simulation model for GPRS

shown in Fig. 5.38 corresponds to the configuration shown in Fig. 5.36. That is, every user stack at the base station also includes a transmission line.

The MAC layer is responsible for multiplexing and demultiplexing of the RLC packets of the different users. At the base station side, this model matches reality. However, at the mobile station side, of course, in reality *no* common MAC layer exists. Instead every mobile station has its own MAC layer which monitors all timeslots and filters out the packets addressed for itself. It can be seen as a *virtual common* MAC layer because it is a distributed MAC layer. This is exactly what the MAC model does. Thus the simulation model corresponds to reality regarding this detail. To provide equal throughput on multiple TCP connections, *fair queuing* is necessary. Every user has its own data and acknowledgment queue. The MAC scheduler then periodically looks at the queue of each user in a round-robin manner, and always takes a single RLC packet out of each queue. Thus every user gets the same amount of bandwidth if the offered load is equal. As in the single-user case, acknowledgment packets have precedence against data packets.

When using the multiple user model, we included the option of a **World Wide Web** (WWW) data source instead of the constant rate source. The model is defined in [126]. It models a typical WWW session of a user. The session arrival process is Poisson modeled. A session consists of a variable number of packet calls. The time between packet calls is exponentially distributed. Each packet call consists of a random number of packets. Both the number of packets in a packet call as well as the time between packets in a packet call are exponentially distributed. See Fig. 5.39 for a sample trace of a WWW session. The exact parameters are given in Table 5.8. The resulting

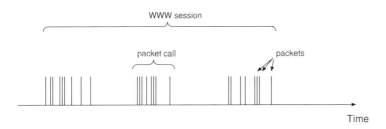

Fig. 5.39. Sample trace of the used WWW packet model

mean data rate *per user* with the model parameters listed in Table 5.8 is about 0.7 kbit/s/user.

To asses the GPRS performance in a realistic environment, we combine circuit- and packet-switched traffic. We used a **model for circuit-switched data** which has the following parameters: The probability of a speech user was 90% and that of a circuit-switched data user 10%. We assumed a Possiondistributed call arrival process. The call arrival process is modeled according to the Erlang B formula. A traffic of 20 mErlang per user and a total number of 160 users for a single carrier frequency is assumed. This leads to a blocking probability of 3%. The mean call duration was also modeled by a Possion process, but different mean values for speech and data traffic were assumed ($1/\mu_{\text{Speech}} = 120\,\text{s}$ and $1/\mu_{\text{Data}} = 300\,\text{s}$). The circuit-switched traffic model is based on empirical values observed in a real GSM network.

Each circuit-switched user occupies a timeslot for the entire time of its activity (call duration). This timeslot cannot be used for packet-switched data any longer. Circuit-switched data always has priority over packet-switched data.

We summarize the default simulation parameters in Table 5.7 and those of the different data sources in Table 5.8.

Table 5.7. Default values of the most important simulation parameters for GPRS simulation

Parameter	Default value
General parameters	
Number of users	variable
TCP	
Version	Reno
Header length	20 bytes
IP	
Maximum IP packet size	65 536
Header length	20 bytes
SNDCP	
TCP/IP header compression	no
Data compression	no
Header length	5 bytes
LLC	
Maximum LLC packet size	1500 bytes
Mode	unacknowledged
Header length	3 bytes
Check sum length	3 bytes
RLC	
Mode	acknowledged/unacknowledged
Packet size	fixed, according channel coding scheme (Table 3.7)
Header length	3 bytes
GSM physical layer	
Environment	'Typical urban'
MS velocity	variable
C/I	variable
SNR	100 dB ($\Rightarrow$ Interference limited)
Forward channel	COST 207 model
Backward channel	COST 207 model
Number of PDCHs	8
MS receiver multislot capability (RxTS)	variable (1..8)

Table 5.8. Default values of the different data sources simulation parameters for GPRS simulation

Parameter	Default value
Constant rate data source	
Packet size	536 bytes
WWW data source	
Packet size	Pareto distributed (k=81.5, α=1.1, mean packet size: 480 bytes)
Maximum packet size	65 535
Session arrival process	Poisson distributed (mean 200 s)
Packet calls per session	exponential distributed (mean 5)
Time between packet calls	exponential distributed (mean 90 s)
Packets per packet call	exponential distributed (mean 25)
Time between packets	exponential distributed (mean 0.125 s)
Final mean data rate per user	about 0.7 kbit/s
Circuit-switched traffic (Speech and data according Erlang B)	
Number of circuit switched users	160
Traffic per user	20 mErlang
Probability of speech users	0.9
Mean call duration for speech users	Poisson distributed (mean 120 s)
Mean call duration for data users	Poisson distributed (mean 300 s)

5.5.2 Simulation Results for GSM-GPRS

In this section we will present simulation results which we obtained by the model presented in the previous section. As there exist many parameters, we investigated only a fraction of the possible parameter combinations, which we thought were important. All simulation results have been made for 'Typical Urban' channel model. A mobile station velocity of $v = 10\,\text{km/h}$ was assumed.

The results can be classified into *single-user* and *multiple-user* results. With the single-user model, it is possible to determine the maximum possible end-to-end performance (throughput and delay) a user can achieve under optimal conditions. Optimal in this sense means that this user can use all available resource exclusively. With the multiple-user model, overall system performance can be investigated. A measure for performance is, for example, the *cumulated total system throughput for all users*.

We will start with the single-user case.

Single-user, net throughput on TCP level, RLC acknowledged mode. We will present the simulated end-to-end mean net throughput experienced by the TCP user for RLC acknowledged mode and a single user in Fig. 5.40. The throughput is plotted over C/I. We plotted all four coding schemes in every figure. The plots differ in the number of timeslots used in parallel for downlink data transmission ((a) RxTS=1, (b) RxTS=2, (c) RxTS=4 and (d) RxTS=8). Figure 5.40(a) shows essentially the possible throughput of each coding scheme per timeslot because RxTS=1 (i.e. only a single timeslot is used).

The nominal throughput values are listed in Table 5.9 (see also Table 3.6). For coding scheme 1 (CS-1) a maximum throughput of 6.5 kbit/s, for CS-2 9.8 kbit/s and for CS-3 10.7 kbit/s was achieved. The mean throughput for CS-4 is about 3.5 kbit/s and much smaller then expected. However, looking at the BLER for this scenario, this is not very surprising. CS-4 has only a minimum BLER of BLER $= 7.5 \times 10^{-1}$ (see Fig. 3.34), which is obviously insufficient for data transmission with RLC acknowledged mode.

The maximum values for the other RxTS, CS-x combinations are also given in Table 5.9. C/I threshold values for acceptable throughput are approximately equal for all coding schemes. About 70% of the maximum throughput is already achieved with a $C/I = 10\,\text{dB}$. But maximum throughput is possible only for $C/I > 15\,\text{dB}$ for CS-1 and $C/I > 20\text{--}25\,\text{dB}$ for CS-2, CS-3 and CS-4.

Figure 5.41 shows the same results as Fig. 5.40, but in different combinations. Now (Fig. 5.41) every coding scheme is plotted in a different plot. A single plot contains various timeslot combinations for one specific coding scheme. In this combination, an increase in throughput for multislot usage can be observed more conveniently. Increasing the number of timeslots from 1 to 2 almost doubles the net throughput for CS-1, CS-2 and CS-3. This can be observed also for doubling the timeslots from 2 to 4. Again the net throughput has almost doubled. However, when going to 8 timeslots, a degradation

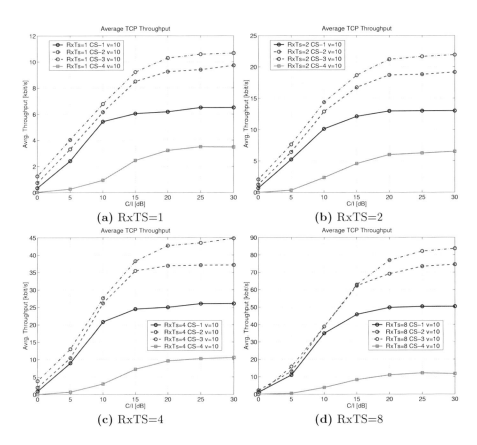

Fig. 5.40. Simulated **TCP mean throughput** of GPRS for a **single user**. In every plot all four coding schemes are shown. The plots differ in the number of MS receiver timeslots, RxTS=1 **(a)**, RxTS=2 **(b)**, RxTS=4 **(c)**, RxTS=8 **(d)**. **RLC acknowledged** mode was used. Mobile station speed was assumed to be $v = 10$ km/h

against the expected throughput occurs in all coding schemes. The degradation is not very high, but still surprising.

A possible cause for this effect is that the window size of the RLC protocol, fixed at 64 packets, is too small for all 8 timeslots to operate in parallel. A further indication which strengthens this conjecture is given by the plots of the MAC data buffer in Fig. 5.45. See Sect. 5.5.3 for further discussion on this topic.

Throughput results from our simulations are comparable with those in [69] at least for CS-1, CS-2 and CS-3 coding schemes: [69] obtains a maximum RLC/MAC layer throughput of 7.2 kbit/s for TU3 channel model with no frequency hopping for CS-1. The channel model TU3 corresponds to 'Typical

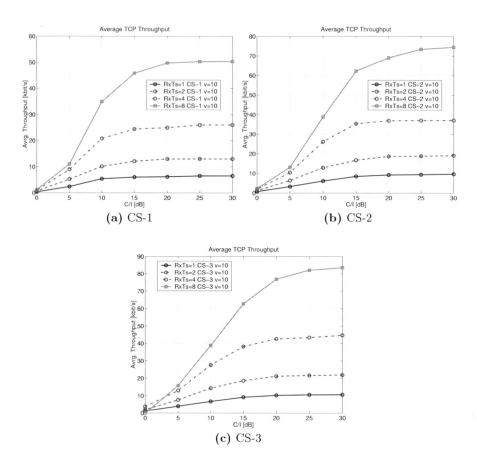

Fig. 5.41. Simulated **TCP mean throughput** of GPRS for a **single user**. The plots differ in the coding scheme used, CS-1 **(a)**, CS-2 **(b)**, CS-3 **(c)**. The number of MS receiver timeslots, RxTS, was varied in every plot. **RLC acknowledged mode** was used. Mobile station speed was assumed to be $v = 10$ km/h

Urban' with 3 km/h mobile station velocity. Note that this model is different from the COST channel model which we used in our simulations.

We obtained a maximum net throughput at TCP layer of 6.5 kbit/s for CS-1. With the data packet size of 536 bytes, this corresponds to 1.516 packets/s. Adding the overhead for TCP, IP, SNDCP and LLC layer leads to an overhead rate, η_o, of,

$$\text{Overhead per data packet} = 20 + 20 + 5 + 3 = 48 \text{ bytes} . \tag{5.23}$$

Overhead rate:

$$\eta_o = 1.516 \text{ data packets/s} \times 48 \text{ bytes/data packet} = 0.58 \text{ kbit/s} . \tag{5.24}$$

Simulated net throughput plus the overhead rate results in an RLC layer throughput of

$$6.5\,\text{kbit/s} + 0.58\,\text{kbit/s} = 7.08\,\text{kbit/s} . \tag{5.25}$$

This is slightly below the RLC layer throughput reported in [69] (7.2 kbit/s).

A special, expected behavior of the four coding schemes is *not* observed in Fig. 5.40. As the coding is best for CS-1 and least powerful for CS-3 (CS-4 has no coding at all), CS-1 should perform best for the lowest C/I value, CS-2 in an intermediate C/I range and CS-3 again in a intermediate, but higher C/I range. CS-4 should give the best results of all coding schemes for very high C/I. This behavior has been published previously [69, 83]. Reference [69] shows only RLC level throughput simulations, whereas in [83] TCP throughput performance is given.

However, we did *not* obtain this result in our simulations. Throughput of CS-4 is inferior and throughput of CS-3 is *always* higher than for CS-2 for the entire C/I range. Also CS-2 performance is always better than CS-1. *No crossing of throughput curves can be observed!*

These astonishing result might be caused by the different channel models used in both simulations. Also CS-4 seems to need a better receiver (demodulator) or a more friendly radio environment to achieve at least a better performance than CS-3.

The nominal and maximum observed throughput as well as the relative degradation for all coding schemes and multislot capabilities is listed in Table 5.9. Throughput performance is always about 70% below the maximum throughput. This is caused by the overhead introduced by the GPRS protocol stack. For 8 timeslots the relative throughput is slightly below 70%. We already discussed this above (see also Sect. 5.5.3).

Table 5.9. Nominal and simulated net throughput for different number of allocated timeslots for all four coding schemes. Also the relative throughput to the nominal one is shown

| Coding scheme | Nominal and simulated throughput (kbit/s) | | | | | |
| | 1 TS | | | 2 TS | | |
	nom.	max.	rel.	nom.	max.	rel.
CS-1	9.05	6.5	72%	18.1	13	72%
CS-2	13.4	9.8	73%	26.8	19	71%
CS-3	15.6	10.7	68%	31.2	22	70%
CS-4	21.4	3.5	16%	42.8	7	16%
	4 TS			8 TS		
	nom.	max.	rel.	nom.	max.	rel.
CS-1	36.2	26	71%	72.4	50	79%
CS-2	53.6	37	69%	107.2	84	69%
CS-3	62.4	45	72%	124.8	84	67%
CS-4	85.6	11	12.8%	171.2	11	6.4%

Single-user, delay on TCP level, RLC acknowledged mode. The end-to-end delay of TCP packets is plotted in Fig. 5.42. The end-to-end delay includes also the delay of the external network, beyond the mobile network. Thus the results in this section are not directly comparable to the definition of the packet delay in [61]. There, only the delay within the mobile network is specified. The delay is plotted as a function of C/I. We plotted four different timeslot combinations in every plot. The plots differ in the channel coding scheme ((a) CS-1, (b) CS-2, (c) CS-3 and (d) CS-4).

In general, TCP delay decreases with the number of timeslots used. Also it decreases when using a coding scheme with higher bandwidth (except CS-4 which has unacceptable performance).

CS-1 shows the same behavior as that observed in GSM-CSD: Delay is high for very poor radio conditions and decreases with better C/I. For low

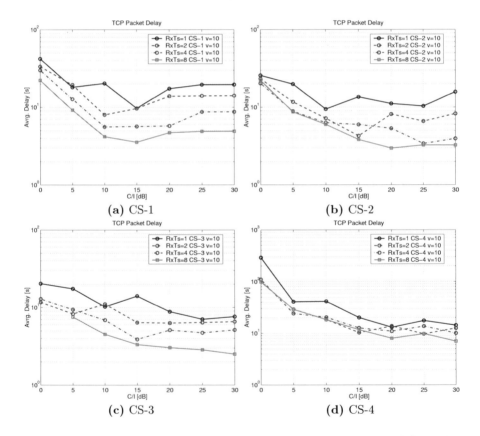

Fig. 5.42. Simulated **TCP mean delay** for a **single user**. The plots differ in the coding scheme used, CS-1 **(a)**, CS-2 **(b)**, CS-3 **(c)**, CS-4 **(d)**. The number of MS receiver timeslots, RxTS, was varied in every plot. **RLC acknowledged** mode was used. Mobile station speed was assumed to be $v = 10\,\mathrm{km/h}$

C/I the delay is introduced by the high number of errors and huge number of necessary RLC retransmissions. However, at some point (for CS-1: $C/I \approx$ 15 dB) the channel quality is good enough that intermediate buffers can fill up. This lets the end-to-end delay increase again. This effect is not very pronounced for CS-2 and CS-3 channel coding schemes, because the coding is worse, and thus the C/I values for minimal delay are shifted to higher C/I values. In stationary state for very high C/I, the TCP delay ranges from 5 s to 20 s (CS-1, RxTS=8..1, respectively). For CS-2 the delay is lower and ranges from 3 s to 15 s, while CS-3 gives lowest TCP delay ranging from 2.5 s up to 8 s.

As already mentioned in the GSM-CSD section, the delay for high C/I is mainly introduced by filled buffers and will be significantly smaller when using interactive traffic.

As an example, the TCP delay distributions for CS-1 and $C/I = 10, 15, 20$ dB are shown in Fig. 5.43. This shows the distributions of the mentioned C/I range from Fig. 5.42(a). The CDFs have the same shape and are shifted in C/I according to their mean delay.

But an exception for 8 parallel used timeslots exists (Fig. 5.43(d)): For high $C/I = 20$ dB almost all TCP packets experience a delay of slightly below 5 s. If 8 timeslots are available, the bandwidth for TCP is high enough to get an almost stable congestion window. Thus the delay of the packets is almost equal. This behavior is close to that of an ideal link. In an ideal link all packets have the same delay in stable state (after slow start).

Single-user, results at RLC level, RLC acknowledged mode. In this section we will present results from the RLC and MAC layer entities. Again, these results are for the single-user case and RLC acknowledged mode.

The first figure (Fig. 5.44) shows the mean RLC packet delay. RLC packet delay is measured as the time a packet needs from RLC sender to the RLC receiver. This includes transmission delay *and* MAC queueing delay. We plotted all four coding schemes in every plot. The plots differ in the number of timeslots. Note the different scaling in Fig. 5.44(a) compared to the 3 other plots.

In general, RLC delay decreases with increasing number of used timeslots. Also, it decreases for increasing C/I. The RLC packet delay is highly correlated with the MAC buffer occupancy and the number of RLC retransmissions. Thus we also show the internal values: Figure 5.45 shows the mean MAC *data* buffer occupancy versus C/I. The sub-plot parameters and partitioning corresponds to that used also in the previous figures. Figure 5.46 shows the mean number of RLC retransmissions versus C/I. Again all parameters are equal to the previous figures.

A number of different effects work together to form the actual RLC packet delay. Every effect can be seen in different regions of the plots in Fig. 5.44. We will discuss each effect separately:

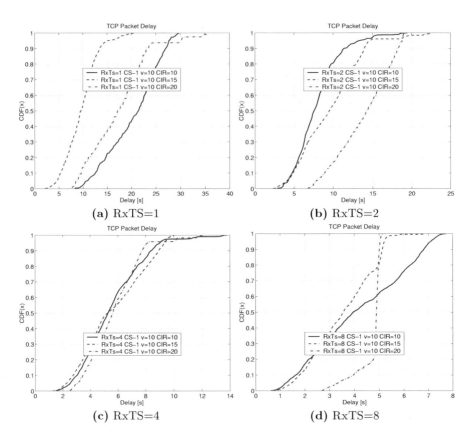

Fig. 5.43. CDF of simulated **TCP delay** for a **single user** for coding scheme CS-1. The plots differ in the number of MS receiver timeslots, RxTS=1 **(a)**, RxTS=2 **(b)**, RxTS=4 **(c)**, RxTS=8 **(d)**. The C/I was varied from 10 dB to 20 dB. **RLC acknowledged** mode was used. Mobile station speed was assumed to be $v = 10$ km/h

1. **Convergence of RLC delay at high C/I:**
   For very high C/I values the three well-operating coding schemes (CS-1, CS-2, CS-3) converge to equal delays. The radio link produces almost no errors, thus the number of retransmissions is small. Therefore the delay (consisting of MAC queueing and transmission delay) only depends on the MAC queuing size which is now *only* fed from real data packets. Queue size for high C/I tends to be at a maximum queue size of 64 packets (see Fig 5.45, the reason for this is explained later). Therefore, if a new RLC packet arrives at the top of the queue, almost always 63 packets are in the queue. The packet has to wait 63 packet transmission time periods until it is transmitted assuming only one available times-

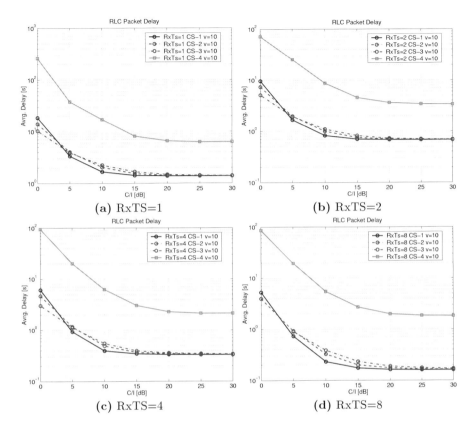

Fig. 5.44. Simulated **delay of RLC packets** for a **single user**. The plots differ in the number of MS receiver timeslots, RxTS=1 **(a)**, RxTS=2 **(b)**, RxTS=4 **(c)**, RxTS=8 **(d)**. All four coding schemes are plotted in every plot. **RLC acknowledged** mode was used. Mobile station speed was assumed to be $v = 10$ km/h

lot[12]. However, a packet transmission time period is always 4 TDMA frames (4×4.616 ms=18.464 ms). This is independent of the channel coding scheme.

Thus, delay is equal for all channel coding schemes for a specific number of timeslots, and the curves converge for high C/I values.

2. **Absolute difference of RLC delay for different RxTS values**:
   The absolute difference in RLC delay for RxTS=1, RxTS=2, RxTS=4 and RxTS=8 can be observed best for high C/I values in Fig. 5.44. It is the difference from sub-plots (a), (b), (c) and (d) on the right-hand side of each plot.

---

[12] This is only valid when we ignore idle frames and other logical channels, like PRACH, which consume additional resources and cause additional delays.

The difference corresponds to the average waiting time of RLC packets in the MAC buffer. As mentioned in *item 1*, MAC buffer has high probability that it is fully loaded to 64 packets. If only a single timeslot is available for transmission, the waiting time for every packet in the buffer can be computed as $4 \times 4.616\,\mathrm{ms} \times 64 = 1.18\,\mathrm{s}$, i.e. 64 block transmission periods, $\tau_{\mathrm{transm}} = 4 \times 4.616\,\mathrm{ms} = 18.464\,\mathrm{ms}$, until this packet is transmitted. Then the actual transmission time τ_{transm} must be added which gives us the total RLC packet delay for RxTS=1 to $\tau_{\mathrm{RLC}} = 1.20\,\mathrm{ms}$. This value can be found almost exactly in Fig. 5.44(a).

When going to multislot operation, MAC buffer waiting time is reduced appropriately, Because multiple packets can be transmitted in parallel. It has the same effect if the MAC buffer size were to be reduced (divided) according to the number of multiple slots used. Now we can compute the minimal RLC delay for every multislot combination easily. The results are shown in Table 5.10.

Table 5.10. Minimum RLC delays for different number of multiple timeslot operations. The values can be observed in Fig. 5.44 and are valid for all coding schemes

Number of timeslots RxTS	MAC buffer length N_{MAC}	MAC waiting time (s) $\tau_{\mathrm{MAC}} = \frac{N_{\mathrm{MAC}} \cdot \tau_{\mathrm{transm}}}{\mathrm{RxTS}}$	Overall RLC delay (s) $\tau_{\mathrm{RLC}} = \tau_{\mathrm{MAC}} + \tau_{\mathrm{transm}}$
1	64	1.18	1.20
2	64	0.59	0.61
4	64	0.29	0.31
8	64	0.147	0.166

3. **Increase in RLC delay at low C/I values:**
   RLC delay increases significantly at low C/I values. For example with a single available timeslot, RLC increases up to $12\,\mathrm{s}$ for $C/I = 0\,\mathrm{dB}$ compared to $1.2\,\mathrm{s}$ at high $C/I > 20\,\mathrm{dB}$ (see Fig. 5.44(a)).

   This effect is caused by the growing number of retransmissions for low C/I values. Figure 5.46 shows the mean number of RLC retransmissions versus C/I. The number of retransmission varies for the different coding schemes, but does *not* vary for different multislot capabilities. Thus the four sub-plots in Fig. 5.46 are *almost equal*.

   Looking at $C/I < 10\,\mathrm{dB}$ the number of retransmissions grows exponentially up to 10 retransmissions for every sent packet. Thus the MAC data queue gets overwhelmed by retransmission packets, and the actual data packets have to wait accordingly. This figure of 10 retransmissions corresponds to the increase in delay for $C/I = 0\,\mathrm{dB}$ of almost exactly one order of magnitude compared to the minimum delay.

4. **Relative difference of RLC delay between coding schemes at low C/I:**

The RLC delay at low C/I values differs slightly for each coding scheme, whereas they converge at higher C/I values (see *item 1*). This effect is caused by the different coding power of the channel codes. CS-1 has the most powerful code, thus smallest number of errors and subsequently the smallest number of retransmitted packets (see Fig. 5.46). CS-4 includes no coding, thus the number of retransmissions is highest. For high C/I values ($C/I > 15$–$20\,\mathrm{dB}$) the mean number of retransmissions is below 10^{-1} for every packet. This low value does not significantly influence RLC delay (via the more heavily occupied MAC buffer). Thus the differences between the coding schemes are also not significant.

But, at low C/I values, retransmissions of packets *do* have a significant influence on RLC delay. As the retransmissions differ for the coding schemes, also RLC delay differs between the coding scheme. This is what we observe in Fig. 5.44 for low C/I values.

Also, the crossovers between the coding scheme curves in both Fig. 5.44 and Fig 5.46 correspond exactly.

Mean MAC data buffer occupancy is shown in Fig. 5.45. Again all four coding schemes are plotted in every sub-plot.

In general, buffer occupancy increases with decreasing channel error rate (increasing C/I). However, this is not true for low multislot capability: When only a single timeslot is available (RxTS=1), the MAC buffer load remains at a constant mean level (≈ 40 RLC packets) below $C/I \leq 10\,\mathrm{dB}$. This implies that the channel quality (with channel coding) is good enough that the TCP sender's congestion window remains sufficiently open. Thus there are always packets available at the TCP sender foreseen to be sent immediately. This effect of settling the buffer occupancy at a bad channel quality diminishes with increasing number of available timeslots. For RxTS=2 (Fig. 5.45(b)), the settling occurs at a lower level (≈ 30 RLC packets). Additionally, the buffer load is no longer constant. For more than 2 timeslots the effect has already totally diminished. Thus, when the bandwidth of the channel is larger, more packets are transmitted over the channel simultaneously. However, when C/I is low, more packets can also get corrupted. Thus, more RLC retransmissions are also necessary, and the experienced delay at TCP level grows. As a consequence, the TCP retransmission timer expires. This causes the TCP sender to reduce the congestion window, thus less TCP packets will be sent.

When the channel quality is very good (high C/I values), the MAC buffer gets loaded up to 63 packets. This can be explained easily by the RLC transmission window size, k. k is fixed at $k = 64$, independent of the channel coding scheme. k is measured in packets, thus the real buffer space in bytes depends on the channel coding scheme. So the RLC sender is able to have 64 packets in the transmission line[13]. Therefore the MAC buffer can be loaded

---

[13] The transmission line for the RLC sender includes the actual channel but also the MAC layer including the MAC buffers.

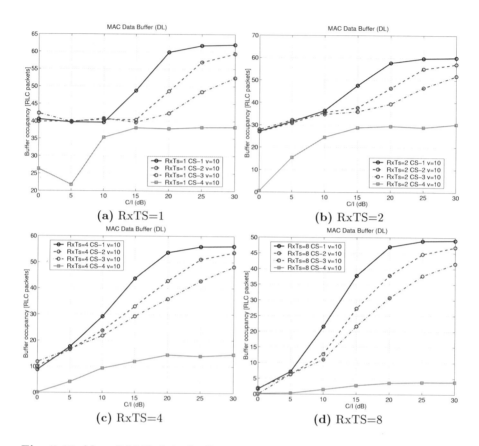

Fig. 5.45. Mean **MAC data buffer occupancy** measured in RLC packets for a **single user**. The plots differ in the number of MS receiver timeslots, RxTS=1 **(a)**, RxTS=2 **(b)**, RxTS=4 **(c)**, RxTS=8 **(d)**. All four coding schemes are plotted in every plot. **RLC acknowledged** mode was used. Mobile station speed was assumed to be $v = 10\,\text{km/h}$

up to 64 packets at maximum. In reality this value will not be reached exactly, because there will always be at least one packet in the channel.

Single-user, RLC *unacknowledged* mode. Now we turn to RLC **un**acknowledged mode with a single user. We will present TCP end-to-end throughput and delay performance.

The mean TCP throughput for all coding scheme–multislot combinations is plotted in Fig. 5.47. The plots differ in the multislot capability in the downlink path.

In unacknowledged RLC mode, TCP is the only protocol providing reliable transmission. Thus TCP needs to make retransmissions in case of errors.

The mean throughput is degraded significantly compared to RLC acknowledged mode. However, maximum throughput is maintained, at least for CS-1

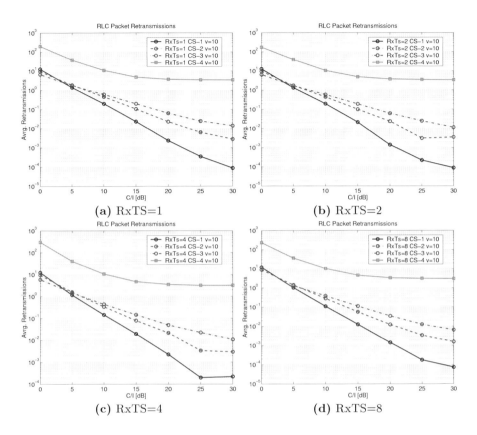

Fig. 5.46. Mean number of **retransmissions of RLC packets** for a **single user**. The plots differ in the number of MS receiver timeslots, RxTS=1 **(a)**, RxTS=2 **(b)**, RxTS=4 **(c)**, RxTS=8 **(d)**. All four coding schemes are plotted in every plot. **RLC acknowledged** mode was used. Mobile station speed was assumed to be $v = 10km/h$.

(see Table 5.9). With CS-1 the maximum throughput is reached at an C/I of 25 dB, which is about 10 dB more in C/I requirement than in RLC acknowledged mode.

The slope of the throughput rises when going to higher C/I. This means that the required C/I threshold for good performance is more stringent than for RLC acknowledged mode.

The next, less powerful, coding schemes, CS-2 and CS-3 do *not* even reach a settled maximum throughput within the simulated C/I range. However, as we know from the link level simulations, BLER will not be significantly lower for even higher C/I values. Thus the throughput at $C/I = 30$ dB should be close to the maximum possible throughput. The maximum mean throughput for all coding schemes in RLC unacknowledged mode is summarized in Table 5.11.

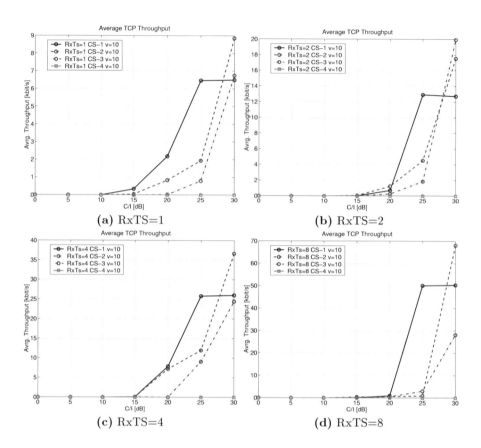

Fig. 5.47. Simulated **TCP mean throughput** of GPRS for a **single user** in RLC *unacknowledged* mode. In every plot all four coding schemes are shown. The plots differ in the number of MS receiver timeslots, RxTS=1 **(a)**, RxTS=2 **(b)**, RxTS=4 **(c)**. Mobile station speed was assumed to be $v = 10$ km/h

Maximum performance for CS-1 is equal to acknowledged mode at 10 dB higher required C/I. CS-2 has at least similar maximum performance to acknowledged mode, but again about 10 dB more in C/I is required. CS-3 performs very poor, throughput is even below the one with CS-1 coding scheme. CS-4 yielded no measurable throughput at all. CS-2 and CS-3 schemes require at least $C/I > 25$ dB.

The end-to-end TCP delay is plotted in Fig. 5.48. TCP packet delay is generally smaller compared to RLC acknowledged mode. Dependent on the multislot capability, the mean delay at $C/I = 30$ dB (maximum throughput, RxTS=8 to RxTS=1) is 3 s to 23 s for CS-1. The relatively low delay for CS-2 and CS-3 from about 0.5 s to 4 s indicates that intermediate buffers are not loaded and TCP's congestion window is not in a stable state.

The RLC packet delay is plotted in Fig. 5.49.

Table 5.11. Nominal and simulated net throughput for different number of allocated timeslots for all four coding schemes and RLC **un**acknowledged mode. Also the relative throughput to the nominal one is shown

| Coding scheme | Nominal and simulated throughput (kbit/s) | | | | | |
| | 1 TS | | | 2 TS | | |
	nom.	max.	rel.	nom.	max.	rel.
CS-1	9.05	6.5	72%	18.1	13	72%
CS-2	13.4	8.9	66%	26.8	17.5	65%
CS-3	15.6	6.8	43%	31.2	20	64%
CS-4	21.4	–	–	42.8	–	–
	4 TS			8 TS		
	nom.	max.	rel.	nom.	max.	rel.
CS-1	36.2	26	71%	72.4	50	79%
CS-2	53.6	37	69%	107.2	68	63%
CS-3	62.4	34	38%	124.8	28	22%
CS-4	85.6	–	–	171.2	–	–

As unacknowledged mode RLC does *not* perform retransmissions, the RLC delay should be more predictable than in acknowledged mode. However, again the MAC buffer delay is also included. Flow-control mechanisms in RLC unacknowledged mode are equal to those in acknowledged mode. The RLC receiver still sends acknowledgment packets (in unacknowledged mode)! But the RLC sender uses them only to send a new packet and *not* to validate whether the packet has been received properly. Acknowledgment packets therefore are necessary to maintain the flow control.

But, as the flow control also fills up the MAC buffer when operating over a reliable link, RLC delay is equal to the one in acknowledged mode. This can be observed for CS-1 at $C/I \approx 30\,\mathrm{dB}$ in Fig. 5.49. The delays are almost exactly the same than observed in Fig. 5.44 for high C/I. Thus, when RLC in unacknowledged mode is working well, no difference in delay is observed.

But when operating on poor channel conditions, which start rather early in unacknowledged mode, delay *decreases*. This is due to the MAC buffer being not fully loaded, because TCP does not maintain a continuous data flow. In the extreme case that the MAC buffer is free, only the plain packet transmission delay remains ($4 \times 4.616\,\mathrm{ms} = 18.4\,\mathrm{ms}$). This is observed at low C/I values in Fig. 5.49.

Multiple users, WWW traffic model. The extension to multiple users sharing the same radio packet data channels (PDCHs) is important for the assessment of GPRS. From the results of this section we will show overall system throughput analysis for a variable number of users. Every user produces traffic according to the WWW traffic model described in Sect. 5.5.1 (see Table 5.8). The mean generated throughput per user is about 0.7 kbit/s. However, the

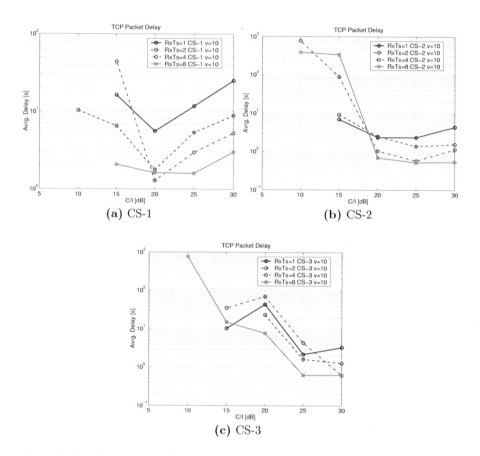

Fig. 5.48. Simulated **TCP mean delay** for a **single user** in **RLC unacknowledged** mode. The plots differ in the coding scheme used, CS-1 (**a**), CS-2 (**b**), CS-3 (**c**), CS-4 (**d**). The number of MS receiver timeslots, RxTS, was varied in every plot. Mobile station speed was assumed to be $v = 10 \, \text{km/h}$

traffic is bursty, therefore a projection to the total system throughput, or the maximum number of allowed users can not be extrapolated easily.

The investigated throughput and delay class were 'best effort'. All users had the same priority.

The simulated traffic time was 45 min. Due to the high computational effort per simulation, and the large memory and storage demands only a single simulation per data point was performed. This explains the somewhat *jerky* shape of the curves. Nevertheless, the main results can be observed clearly. In all simulations we used the RLC acknowledged mode, a C/I of 30 dB and the mobile station velocity was set to $v = 10 \, \text{km/h}$.

Figure 5.50 shows the **total system throughput** on **RLC layer** versus the number of users. The plots differ in the number of available PDCHs

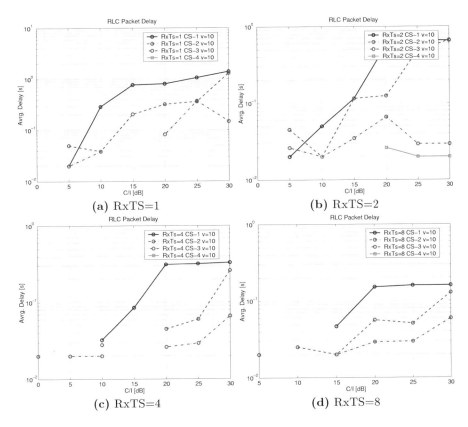

Fig. 5.49. Simulated **delay of RLC packets** for a **single user** in **RLC** *unacknowledged mode*. The plots differ in the number of MS receiver timeslots, RxTS=1 **(a)**, RxTS=2 **(b)**, RxTS=4 **(c)**, RxTS=8 **(d)**. All four coding schemes are plotted in every plot. Mobile station speed was assumed to be $v = 10$ km/h

(PDCH=1,2,4,8). In every plot three different downlink multislot capabilities of the mobile stations (RxTS=1,2,4) are included.

Figure 5.50(a) shows the total RLC system throughput for one PDCH. Increase in throughput is linear with the number of users up to 5–6 users. Then throughput saturates at 6 kbit/s this is only marginally below the throughput for the single user case (see Fig. 5.40(a), CS-1, RxTS=1). The saturation point is independent of the multislot capabilities (RxTS) of the mobile stations. This is not very surprising, because only a single PDCH is available and thus a higher multislot capability does not help.

If more PDCHs are available (Figs. 5.50(b,c,d)), the saturation point in terms of the number of users scales almost linearly with the PDCHs (e.g. about 10 users for 2 PDCHs, and about 20 users for 4 PDCHs). Also the maximum total system throughput scales linearly from about 12.5 kbit/s for

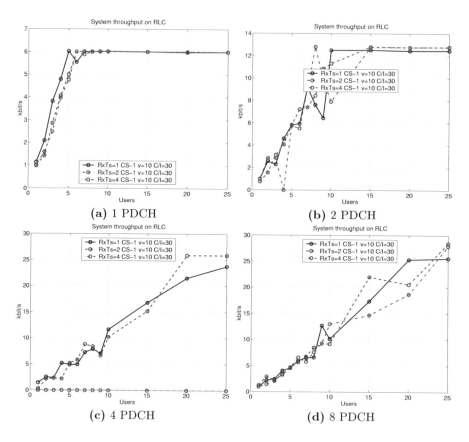

(a) 1 PDCH   **(b)** 2 PDCH

(c) 4 PDCH   **(d)** 8 PDCH

Fig. 5.50. Simulated **total system throughput** on **RLC layer** in **RLC ac-knowledged** mode. Each user sends data according to the WWW model in Table 5.8. The plots differ in the number of available packet data channels (PDCH), 1 PDCH **(a)**, 2 PDCH **(b)**, 4 PDCH **(c)** and 8 PDCH **(d)**. The number of parallel timeslots possible for one user is the parameter RxTS=1,2,4. Mobile station speed was assumed to be $v = 10$ km/h and coding scheme CS-1 was used

two PDCH up to 26 kbit/s for 4 PDCHs. The range of simulated users is too small to observe the saturation point for 8 PDCHs, but a linear extrapolation should lead to correct results.

In every plot, three multislot capabilities are displayed simultaneously (RxTS=1,2,4). However, there is *no* significant difference between them. This is observed regardless of the number of available PDCHs. Thus, multislot capability does *not* increase the total system throughput. **From an operators point of view, the multislot class does not affect the system throughput.** However, the individual user does gain in his throughput performance.

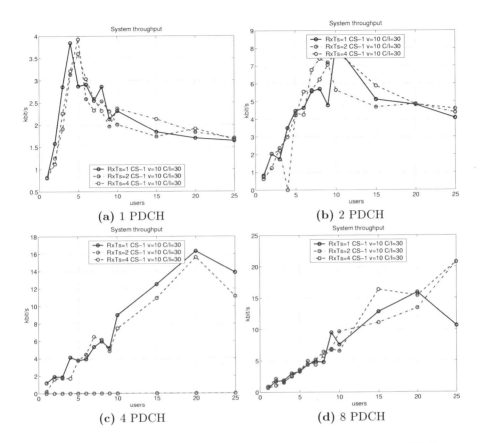

Fig. 5.51. Simulated **total system throughput** on **TCP layer** in **RLC acknowledged** mode. Each user sends data according to the WWW model in Table 5.8. The plots differ in the number of available packet data channels (PDCH), 1 PDCH **(a)**, 2 PDCH **(b)**, 4 PDCH **(c)** and 8 PDCH **(d)**. The number of parallel timeslots possible for one user is the parameter RxTS=1,2,4. Mobile station speed was assumed to be $v = 10$ km/h and coding scheme CS-1 was used

The total system throughput on **TCP level** is plotted in Fig. 5.51. As in the previous figure, the plots differ in the number of available PDCHs (PDCH=1,2,4,8) and every plot includes all 3 investigated multislot capabilities (RxTS=1,2,4).

Considering a single PDCH (Fig. 5.51(a)) the total system throughput increases linearly with the number of users up to 5 users. There, the TCP system throughput is slightly below 4 kbit/s. This maximum corresponds to the saturation point observed in the previous figure for the RLC system throughput. On increasing the number of users beyond this point, the total TCP system throughput *falls down* again. This decrease is exponential with a growing number of users, and it saturates at about 43% of the maximum system throughput.

The same behavior is observed for more available PDCHs (Figs. 5.51(b,c,d)). Only the number of users where the maximum TCP system throughput is observed is shifted according to the available PDCHs. Again, maximum throughput scales linearly with the available PDCHs, from 4 kbit/s to 16 kbit/s for 1 to 4 PDCHs. The maximum TCP throughput is about 30% below the RLC system throughput.

As for the RLC system throughput, the multislot capability of the mobile stations has *no* influence on the TCP system throughput.

The fact of decreasing TCP system throughput when the number of users is beyond a critical number is very important. Although RLC throughput is stable beyond this point, TCP throughput performance drops. Thus the individual end user experiences a TCP performance that is *below* the linearly down-scaled performance. Linearly down-scaled TCP performance for increasing number of users would imply that the total system TCP throughput beyond the critical point is *stable* at the maximum possible one. But this is not the case, as identified previously.

When more users than the critical number are attached to the system, TCP is not able to acquire the bandwidth needed according to the offered load from the traffic model. Thus the congestion window is very small and TCP is always in 'slow start' mode. In 'slow start' TCP tries to increase the congestion window exponentially, but fails, often due to the heavy congestion. This leads to frequent TCP timeouts and significant performance decreases.

Note an important point considering the absolute values observed: The critical number of GPRS users depends on the specific WWW traffic model. The evaluated numbers of users in this section are *only* valid for this particular traffic model. Another traffic model might lead to completely different results. However, the throughput values will *not* change. Only the user axis will be scaled. Maximum RLC and TCP system throughput depend only on the coding scheme, the number of allocated PDCHs and the radio environment.

Multiple users, combined packet- and circuit-switched traffic. To assess the performance of a GPRS system under realistic environments it is necessary to take circuit-switched traffic into account. We modeled circuit-switched speech and data traffic according to the Erlang B formula. We assumed 160 circuit-switched users with a mean Erlang traffic of 20 mErlang per user. Remember, we model only a single carrier frequency. Circuit-switched traffic has priority over the packet-switched traffic, thus the effectively available PDCHs are reduced according to the actual circuit-switched traffic. The detailed parameters have been described in Sect. 5.5.1 and are listed in Table 5.8.

Figure 5.52 shows the total packet-switched system throughput on both RLC layer (a) and TCP layer (b). Eight PDCHs have been assigned, i.e. in every timeslot packet data traffic was allowed (if it was not occupied by

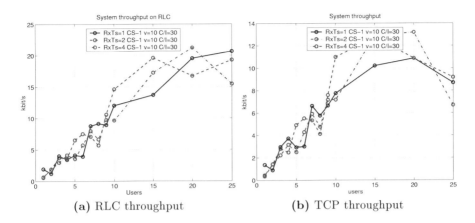

(a) RLC throughput

(b) TCP throughput

Fig. 5.52. Simulated **total system throughput** on **RLC layer (a)** and **TCP layer (b)** in a **combined packet- and circuit-switched** traffic environment. For GPRS user, RLC acknowledged mode was used, and every packet user sends data according to the WWW model in Table 5.8. Eight PDCHs were available for packet data but are shared with circuit-switched speech and data users. The number of parallel timeslots possible for a single data user is varied (RxTS=1,2,4). Mobile station speed was assumed to be $v = 10 \, \text{km/h}$ and coding scheme CS-1 was used

circuit-switched traffic). The multislot capability of the GPRS mobile stations was varied from RxTS=1 to 4. The equivalent figures without mixed traffic are Fig. 5.50(d) for the RLC layer and Fig. 5.51(d) for the TCP layer throughput. Note the different scalings of the ordinate.

Up to about 15 WWW users, no significant difference in RLC layer throughput can be observed. The additional circuit-switched traffic does not have a measurable influence on total RLC system throughput. If more than 15 WWW users are active, the system RLC layer throughput saturates at about 20 kbit/s.

At the TCP layer, the influence is even more pronounced. Again, the influence is measurable for more than 15 WWW users. The TCP layer total system throughput has its maximum for 15 to 20 WWW users. Allowing more WWW users in the system leads to a system *overload* which reduces overall TCP throughput. Thus the individual user also observes a lower TCP throughput.

From the observed figures we conclude that additional circuit-switched traffic reduces the overload threshold user number significantly. In this example it was reduced from about 40 WWW users to about 20 WWW users for 8 PDCHs.

5.5.3 Discussion and Conclusions

RLC acknowledged mode issues. Maximum mean net end-to-end throughput in RLC acknowledged mode is, in general, about 70% of the theoretical optimum for CS-1, CS-2 and CS-3. Coding scheme CS-4 throughput performance is inferior and always below *all* other coding schemes. The threshold C/I required to achieve more than 90% of the maximum throughput is for CS-1 at $C/I_{\text{thres}} = 10\text{--}15\,\text{dB}$, whereas for CS-2 and CS-3 it is shifted to $C/I_{\text{thres}} = 20\,\text{dB}$.

Considering the net TCP throughput for different numbers of parallel timeslots, the combination of plots in Fig. 5.41 gives a convenient view of the effect of multislot usage. Increasing the number of timeslots gives the expected growth in throughput (i.e. doubling the timeslots from 1 to 2 doubles also the throughput). However, this is *not* true when going to the maximum possible number of parallel timeslots (8). A degradation against the expected throughput can be observed.

Also it is interesting that this effect is the same for all coding schemes. When 8 timeslots are used, the throughput does *not* scale according to the allocated timeslots.

Why is the RLC acknowledged mode protocol *not* suitable for data transmission at maximum possible rate (8 timeslots)?

Three observations are of interest:

1. The throughput scales according to the number of timeslots used, but *not* when going to 8 timeslots any more. We have already mentioned this.
2. This effect (item 1) is similar in all four coding schemes though the absolute throughput values are different. Because RLC counts all state variables in packets and not in bytes (as TCP for example), the actual coding scheme has *no* influence on the proportion between number of RLC packets and number of timeslots. The number of RLC packets in transit is similar in all coding schemes[14]. Therefore, the effect of the RLC protocol should be equal to all coding schemes.
3. The MAC buffer occupancy *does not* reach the RLC window size when 8 timeslots are used. From 1 to 4 available timeslots, the MAC buffer has a mean size of 63 to 55 RLC packets, according to 100% to 87% of the maximum of 63 packets (see Fig. 5.45). However, when using 8 timeslots, MAC queue length is only filled up to 76% of the maximum possible length. Thus the link may not be fully utilized all the time.

We conclude:
Net TCP throughput scales with the number of timeslots used. But, the RLC acknowledged mode transmission window size ($k=64$) might

---

[14] This is also why RLC can operate with all coding schemes with the same window size. The window size is measured in packets, and the packets have different lengths in each coding scheme.

not be suitable for a transmission over all 8 timeslots in parallel. The transmission bandwidth seems to be not fully utilized, therefore the maximum throughput can *not* be reached. Independent of the number of timeslots, coding scheme CS-4 is unsuitable for acknowledged mode.

Acknowledged vs. unacknowledged RLC mode. Comparing the mean net TCP throughput in the RLC acknowledged (Fig. 5.40) and RLC unacknowledged modes (Fig. 5.47), the following observations are important:

- Maximum mean throughput is equal for CS-1, 71% to 79% of the theoretical optimum. Throughput for CS-2, unacknowledged mode is only slightly below acknowledged mode throughput.
- CS-3 coding scheme provides good performance in acknowledged mode, but inferior performance in unacknowledged mode.
- CS-4 coding scheme performs worse than all other coding schemes in acknowledged mode. In unacknowledged mode even *no* throughput was measurable.
- Threshold C/I to achieve 90% of maximum mean throughput is shifted from $15\,\mathrm{dB}$ (acknowledged mode) to $25\,\mathrm{dB}$ (unacknowledged mode) for CS-1. CS-2 needs in general $5\,\mathrm{dB}$ more in C/I.
- TCP delay is, in general, lower for RLC unacknowledged mode (when throughput is comparable), but results mainly from not fully loaded intermediate buffers. This is because data flow of TCP packets is not as fluid as in RLC acknowledged mode, because TCP also experiences errors that it has to remedy by retransmissions.
- For comparable throughput, RLC packet delay is almost equal in acknowledged and unacknowledged modes. But going to high error rates (low C/I), RLC delay *increases* in acknowledged mode (because of retransmissions) and *decreases* in unacknowledged mode (because the MAC buffer gets evacuated).

We conclude:

For reliable data transmission with TCP, RLC **acknowledged** mode should be used. Only coding schemes CS-1, CS-2 and CS-3 perform well in the 'Typical Urban' environment. CS-4 performance is below all others, thus we do *not* recommend its usage.

RLC *un*acknowledged mode is *not* suited for reliable data transmission. However, packet delay is, in general, lower than in acknowledged mode. It might give sufficient performance for applications that can cope with relatively high error rates, but need predictable packet delays (real-time applications, voice and video). These applications do not use TCP but a different transport protocol, which might not be reliable.

Comparison to GSM circuit-switched data (CSD). The throughput for CS-1 and RxTS=1 is below the throughput simulated for GSM TCH/F9.6 traffic channel. Remember that the GSM simulations were done for the *transparent* mode (see Fig. 5.15), which yielded a maximum of $\approx 8\,\text{kbit/s}$ at $C/I = 20\,\text{dB}$. GPRS CS-1 yielded only about $6.5\,\text{kbit/s}$ in both acknowledged and unacknowledged mode. Only the C/I threshold is different, $10\,\text{dB}$ for RLC acknowledged and $20\text{–}25\,\text{dB}$ for RLC unacknowledged mode. Thus GPRS results in RLC *un*acknowledged mode are comparable to the results for GSM-CSD transparent mode (TCH/F9.6) in terms of C/I threshold. Note also that the nominal data throughput for GPRS CS-1 is about 0.6kbit/s below that for GSM TCH/F9.6. Thus, when channel coding is powerful and well suited for the channel under consideration, the difference between acknowledged and unacknowledged link level mode is only a few percent in maximum throughput anyway. The necessary C/I threshold depends on the used mode (RLC acknowledged and unacknowledged), and this is the same in GPRS and GSM-CSD.

Multiple users, WWW traffic. The total system throughput on all timeslots of a single carrier increases with increasing number of users linearly. But at a specific, critical number of users, the system is fully loaded and throughput on RLC layer is saturated. The saturation level depends on the number of allocated Packet Data Channels (PDCHs) and also scales linearly. The maximum RLC layer system throughput varies from $6\,\text{kbit/s}$ to $26\,\text{kbit/s}$ for 1 to 4 PDCHs.

The critical number of WWW users depends on the WWW traffic model. The parameters of the model used in this section are listed in Table 5.8. For this particular WWW model the critical number of users varies between 5 to 20 WWW users for 1 to 4 PDCHs.

An important observation is the different behavior of RLC and TCP layer total system throughput when the system is fully loaded or even overloaded: Whereas the RLC layer system throughput remains *stable* at the maximum possible throughput, if the number of users exceeds the critical number, TCP layer system throughput becomes *unstable*. That is, TCP system throughput *decreases* exponentially when the number of users *increases*.

We conclude:
A GPRS system is sensitive to overload and therefore it *should not* be overloaded by TCP traffic. TCP performance (total and of each individual user) decreases in overload conditions exponentially for growing number of users, although RLC total system throughput is stable and at its maximum.

The critical number of users when assuming a typical WWW traffic scenario ranges from 5 to 20 for 1 to 4 available Packet Data Channels (PDCHs).

The total system performance is *independent* of the multislot capability of each individual mobile station. However, the performance of every connection is affected when the system is *not* yet fully loaded. In that case, higher multislot capability of the terminal leads to higher throughput for this specific user.

Multiple users, mixed circuit-switched and WWW traffic. We investigated the influence of additional circuit-switched speech and data traffic on the performance of the packet-switched WWW data users. Eight PD-CHs have been allocated, but circuit-switched traffic had precedence over the packet data traffic.

The behavior of the total switched data traffic does not change, in principle, compared to the isolated data traffic system. Again a *critical* number of WWW users is observed up to which the TCP and RLC system throughput is increasing linearly with the number of WWW users. Beyond this point, the RLC throughput saturates and TCP throughput decreases.

However, the critical number of WWW users drops significantly to about *one half* compared to the isolated data traffic system. This also implies a reduced maximum RLC and TCP system throughput of about one half compared to the WWW only traffic case.

We conclude:
Additional circuit-switched speech and data traffic reduces the critical number of WWW users significantly, where the system becomes unstable for TCP data traffic. The exact reduction of servable WWW users depends on the number of circuit-switched users. For the investigated sample model parameters, the reduction was about *one half*.

Thus, in a combined GPRS system with mixed traffic, circuit-switched traffic has significant influence on the performance of packet-switched TCP data traffic.

Expected GPRS performance in a real-life network. We performed simulations *and* measurements for GSM-CSD but only simulations for GPRS. The main difference is the switching paradigm changing from circuit switched to packet switched. A measurement environment for GPRS which is similar to the GPRS simulation environment can only be setup in a test environment but hardly in a real operating GPRS network. Isolated measurements for a single user will be difficult. Also, the service class used in measurements will strongly affect performance. This is not included in our GPRS simulations as only a single user or equally entitled users are simulated.

As the correspondence between GSM-CSD simulation and measurement is very good we will nevertheless try to make a basic prediction of GPRS performance in a real-life network. But see also the next paragraph for a comparison to measured results in a GPRS network. We use the simulation results obtained in this section for RLC acknowledged mode and a single user as reference performance, namely the plots in Fig. 5.40 for TCP throughput and Fig. 5.42 for TCP delay.

Because we used *no* TCP/IP header compression in our GPRS simulations, and this will normally be switched on in a real-life GPRS network, maximum throughput should be slightly increased compared to our simulation results. The exact figure will depend on the TCP segment size, but for CS-1 it will be in the same order as for TCH/F9.6, about 0.5 kbit/s per timeslot. For the other coding schemes, this will be proportionally higher. However, as we will point out in the UMTS section (Sect. 5.6), a reverse effect affects performance if the radio channel quality is bad and many packets get lost. Then performance with active header compression is below that without header compression. This will likely be the case with GPRS CS-4 and possibly, but hardly, also with CS-3.

Thus maximum TCP throughput performance of a single GPRS user in a hardly loaded network will be in the order of the values from Fig. 5.40 plus about 0.5 kbit/s per timeslot.

A second main point is the delay introduced by the mobile network as well as the external packet data network. We modeled both together with a constant delay of $\tau = 10$ ms. In a real-life network this will never be a constant but will vary in a wide range. This parameter is also strongly influenced by the negotiated delay class and the implementation of the internal mobile network (e.g. queuing strategies). Thus we think it is not possible to make a reasonable prediction about the packet delay in the mobile network as well as in the external packet data network.

However, the behavior of the packet delay regarding buffering and radio link performance is included in the results of our simulations. Thus delay in a real-life network should be varying in a specific range around the results from Fig. 5.42. This specific range depends on the internal and external network delay.

Comparison to measured results in the literature. In [87] GPRS measurement results have been published. Bulk data transfer from a server to the mobile station in a live GPRS network. In the downlink direction 2 timeslots and in the uplink only a single one was configured. The reliability class was 'Unacknowledged GTP and LLC; Acknowledged RLC, Protected data'. This is the same class we used in our GPRS simulations. Thus the results are directly comparable.

In the downlink direction (2 timeslots) a mean throughput of 12.5 kbit/s was measured, but 90% of the tests yielded a throughput of about 20 kbit/s. This was also the maximum measured throughput. This measurement results correspond almost exactly to the simulation results we presented (see Table 5.9 and Fig. 5.41(b)). There we obtained a maximum throughput of about 19 kbit/s.

The robustness of the transmission was good in stationary environments, but during moving connections throughput was very unstable. Also, GPRS had a higher variance in performance compared to circuit-switched connections.

Furthermore, the authors in [87] came to the conclusion that TCP can not cope properly with the characteristics of GPRS data transmission. TCP does unnecessary retransmissions causing a significant throughput degradation.

5.6 TCP over UMTS, Simulations

Simulation results for TCP data transmission over UMTS bearer services that have been performed by [92] will be presented in the following. Again, the simulation environment was 'Ptolemy' [78].

As in the previous subsections, we will start with the simulation model, followed by the results. Conclusions will finish this section.

5.6.1 Simulation Model for TCP/IP over UMTS

The overall structure for the *single-user* model is shown in Fig. 5.53. The simulation model for UMTS packet-switched data looks, at the first glance, equal to the GPRS structure (Fig. 5.36). However, major parts have changed significantly. The new parts are mainly located in the access network including the radio interface. Beginning from the topmost left-hand side of Fig. 5.53: A data source is sending data packets into the TCP sender. The IP sender receives the TCP packets and encapsulates them into IP packets that are sent over a transmission line with 2 Mbit/s bandwidth and a delay of $\tau = 100$ ms. The transmission line models the Internet *and* the core network of the mobile

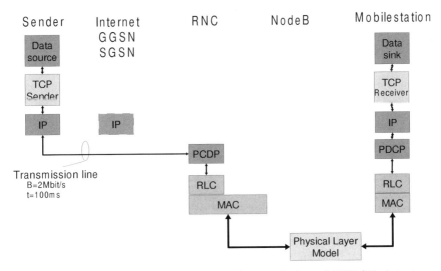

Fig. 5.53. Overall structure of the model for simulation of TCP/IP data transmission over UMTS. For multiple users, the protocol stack entities between data source and RLC layer are duplicated similar to Fig. 5.38

radio system including GGSN and SGSN. At the Radio Network Controller (RNC) the IP packets are packed into PDCP packets, and finally RLC packets are formed and sent via the MAC layer over the channel. At the mobile station packets are decapsulated accordingly in each protocol layer until the data packet is received at the data sink.

Two different kinds of data source models have been implemented. First constant rate source producing equal size packets, which models FTP data transfer. The second source models WWW traffic. The WWW model is simpler than the one used in the GPRS environment. It uses exponential file size distribution with a cut-off at 10 kbyte and 250 kbyte. Only file sizes in this range are transmitted. The delay between files has an exponential distribution between zero and a maximum time (20 s).

TCP version 'Reno' is used in the simulations.

The IP layer frames the TCP packets into IP packets. The IP packets are sent over a transmission line with 2 Mbit/s bandwidth and a delay of $\tau = 100$ ms. This transmission line models the external packet data network (Internet) and the UMTS core network from GGSN, SGSN to the access network (the Radio Network Controller). As the bottleneck is assumed to be at the radio interface, it is reasonable to model the remainder of the connection with a single transmission line.

When the IP packets arrive at the RNC, they are sent to the PDCP layer. The PDCP entitiy adds a PDCP packet identifier header. TCP/IP header compression is supported optionally. The RLC layer model supports *Unacknowledged Mode Data* (UMD) and *Acknowledged Mode Data* (AMD). RLC segments the incoming PDCP packets into small RLC/MAC packets. The size of these small packets depends on the transport format of the radio bearer. For every transport format a specific transport block size is defined. The UMTS physical layer encodes *one* transport block set in a single *Transmission Time Interval* (TTI). A transport block set consists of one or several transport blocks. The transport block size, transport block set size and the TTI for the sample transport formats presented in this book, are summarized in Table 5.12. See also Sect. 4.2 for further explanation of transport blocks.

Table 5.12. UMTS transport formats which are used in this book. TTI is the time between two transport block sets. A transport block set consists of one or several transport blocks (see Sect. 4.2)

Transport format (Bearer service) (kbit/s)	Transport block size (bit)	Transport block set size	Transmission Time Interval (TTI) (ms)
12.2	244	1	20
64	1280	1	20
144	2880	1	20
384	3840	1	10
2048	4096	40	80

The physical layer exchanges an entire transport block set at the same time with the MAC layer as the RLC receiver receives an entire transport block set from the MAC layer. This causes problems at the TCP level when the transport block set is large. See Sect. 5.6.2 for further discussion.

The MAC layer schedules multiple RLC connections over a single radio bearer. Fair scheduling is implemented, i.e. every RLC connection can acquire the same amount of bandwidth. As in the GPRS simulation environment, RLC acknowledgment packets have precedence over RLC data packets.

The physical layer is modeled as *block error model*. We assume uniform distributed block errors. A uniform distribution is reasonable after channel coding and interleaving. No burst errors are considered. A *block* in this context is equal to a transport block. The physical layer delay is determined by the TTI of the transport format.

All important simulation parameters are summarized in Table 5.13.

5.6.2 Simulation Results for TCP/IP over UMTS

Five transport formats have been selected for investigation. They are listed in Table 5.12. We will present the net throughput on the TCP layer for all transport formats in the first part of this section with one exception: The 2 Mbit/s bearer service is investigated separately, because of some important results. The effect of header compression on the throughput is shown afterwards. Finally, we will present FTP and WWW traffic examples.

Sample radio bearers. TCP net throughput for **12.2kbit/s radio bearer** for both RLC acknowledged and unacknowledged mode (RLC-AMD and RLC-UMD) is plotted in Fig. 5.54. Maximum throughput for RLC-AMD is 10.45 kbit/s, for RLC-UMD 11.1 kbit/s which is 85.6% and 90.9% of the theoretical maximum of the radio bearer. Maximum throughput is higher for RLC-UMD because RLC-UMD has less overhead. RLC-AMD performs well up to a BLER of BLER $= 3 \times 10^{-2}$. A higher BLER causes a rapid decrease in TCP throughput. RLC-UMD needs a much better channel and performs well only up to BLER $< 1 \times 10^{-3}$. Thus RLC-UMD needs a BLER better by more than a decade to perform well.

RLC-AMD shows a much steeper slope of decreasing throughput when going to higher BLER. This results from the interaction between RLC-AMD and TCP. As BLER increases, AMD needs to retransmit packets more often. Thus the RLC/MAC queue gets loaded more rapidly, which causes higher delays for all packets. TCP does not know about these higher delays immediately and starts retransmissions because of timeouts. Thus, even more packets have to be transmitted which worsens the situation at the RLC/MAC layer. This interaction is not possible in RLC-UMD, thus throughput decrease is more gentle.

Transport block size is small (244 bits), and comparable to the RLP block size of the GSM TCH/9.6 traffic channel (240 bits). Thus, protocol overhead

Table 5.13. Default values of the most important simulation parameters for UMTS simulation

Parameter	Default value
General parameters	
Number of users	1,2
Constant rate data source	
Packet size	1436 bytes
WWW data source	
File size	exponential distributed (mean = 50 kbyte)
Minimum file size	10 kbyte
Maximum file size	250 kbyte
Time between packets	uniformly distributed (0 to 20 s)
TCP	
Version	Reno
Header length	20 bytes
IP	
Maximum IP packet size	1500
Header length	20 bytes
PDCP	
TCP/IP header compression	Yes/No
Data compression	no
Header length	1 byte
RLC	
Mode	acknowledged/unacknowledged
Packet size	fixed, according to transport format (Table 5.12)
Header length	UMD: 2 bytes/AMD: 4 bytes
MAC	
Scheduling	fair
Header length	4 bits
UMTS physical layer	
Transport format	according to Table 5.12
Channel model	Block error model (uniformly distributed errors)

is relatively high because the RLC entity needs to segment PDCP PDUs into many small blocks.

The performance of the **144-kbit/s radio bearer** is plotted in Fig. 5.55. The maximum observed throughput is 136.8 kbit/s for RLC-AMD and 139.03 kbit/s from RLC-UMD (which is 95% and 96.5% of the raw data rate). This is a much higher relative throughput than on the 12.2-kbit/s radio bearer (95% compared to 85.6%). This is mainly caused by the more than tenfold higher transport block size (2880 bits compared to 244 bits). The much lower overhead leads to higher relative maximum throughput.

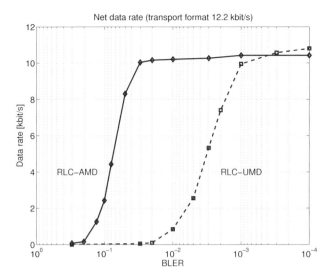

Fig. 5.54. Net TCP throughput for **12.2kbit/s** radio bearer. Both RLC acknowledged and unacknowledged mode (RLC-AMD and RLC-UMD) are plotted

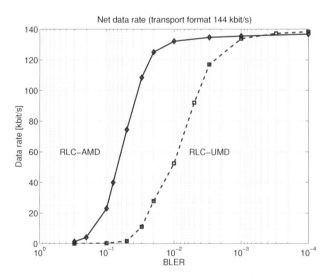

Fig. 5.55. Net TCP throughput for **144-kbit/s** radio bearer. Both RLC acknowledged and unacknowledged mode (RLC-AMD and RLC-UMD) are plotted

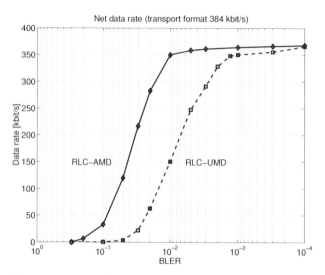

Fig. 5.56. Net TCP throughput for **384-kbit/s** radio bearer. Both RLC acknowledged and unacknowledged mode (RLC-AMD and RLC-UMD) are plotted

The required BLER to achieve 90% of the maximum throughput for RLC-AMD is BLER $= 1.8 \times 10^{-2}$, which is only about one half of the BLER for the 12.2-kbit/s radio bearer, i.e. RLC-AMD performance on the 144-kbit/s radio bearer is slightly *worse*. On the other hand, the required BLER for RLC-UMD is about 2×10^{-3} for $\approx$ 90% of maximum throughput. This is more than the double BLER of the 12.2-kbit/s bearer, i.e. RLC-UMD performs *better* on the 144-kbit/s bearer. The improvement for RLC-UMD is caused by the different transport block size. In the 144-kbit/s radio bearer a TCP/IP/PDCP packet needs to be segmented in only about 1/10 of the RLC blocks required for the 12.2-kbit/s bearer. Thus, fewer segments are discarded at a given BLER. This increases throughput at a given BLER.

However, an important fact has to be taken into account. The channel model is a simple block error model. In a more detailed link level model, BLER at a given *signal-to-noise and interference ratio* (SNIR) will be higher for the 144-kbit/s bearer than for the 12.2-kbit/s radio bearer because of the much larger transport block size[15]. Thus, the above-mentioned effect for RLC-UMD (144-kbit/s bearer performance is much better than 12.2-kbit/s bearer performance) might disappear when replacing BLER with the SNIR at the receiver input as reference dimension.

Throughput performance of the **384-kbit/s radio bearer** for RLC-AMD and RLC-UMD is shown in Fig. 5.56. Maximum throughput is 368.3 kbit/s for RLC-AMD and 371 kbit/s for RLC-UMD (96% and 96.7% of the raw data rate). RLC-UMD performance is similar to that of the 144-kbit/s radio

---

[15] BLER increases with block length when all other parameters remain the same.

bearer. But RLC-AMD performance is worse than for the 144-kbit/s bearer when comparing a specific percentage of maximum throughput.

When comparing the relative performance of TCP data transmission over RLC-AMD mode for all three investigated radio bearers so far (12.2 kbit/s, 144 kbit/s, 384 kbit/s) we can observe a small, but continuous degradation going up to higher data rates. This is caused primarily by the following mechanisms: The number of segments in transit, corresponding to one transmission window, is larger. Thus multiple losses of segments in one transmission window are more likely to happen. This reduces throughput at a specific BLER level. Again be aware that we compare according to BLER here, thus RLC-AMD performance should not change for the different bearers in the ideal case.

Sample radio bearer 2 Mbit/s. The 2-Mbit/s radio bearer needs special consideration. A bandwidth of 2 Mbit/s is achieved by multi-code transmission of 40 transport blocks of size 4096 bits within a TTI of 80 ms. Thus a total transport block set size of 20.48 kbyte is transmitted and received almost simultaneously at the RLC level[16]. This is more than 30% of the maximum standard TCP window size of 65 536 bytes.

At a standard TCP maximum window size of 65 536 bytes a maximum RTT of

$$\mathrm{RTT}_{\max} = \frac{\text{Windowsize}}{\text{Bandwidth}} = 255\,\mathrm{ms} \tag{5.26}$$

is allowed. A MAC buffer size of only 4 transport block set sizes introduces already a delay of 320 ms. Together with the delay of the external packet data network (in this simulation 100 ms), the maximum delay of 255 ms is clearly exceeded. This delay turns the entire network into a *long, fat network* (LFN). Thus, *window scaling* for the 2-Mbit/s bearer is mandatory.

A bursty transport of transport block sets introduces a second problem. TCP is self-clocking and relies heavily on acknowledgments arriving regularly at the sender. These are used for estimation of the RTT. As 40 (!) transport blocks with size 4096 bit arrive simultaneously, this self-clocking mechanism is seriously misled. This leads to erroneous RTT estimation and thus erroneous RTO estimation. Although a single timeout-based retransmission causes the RTO to be doubled, the next 40 transport blocks which arrive simultaneously, decrease the RTO again. Thus RTO is not changed in a large scale. As a final 'result' spurious retransmissions occur which decreases throughput.

In [92] a modification to TCP was suggested to overcome these problems. First, the window scaling factor was set to 2. Secondly, a simple modification to the RTO calculation was accomplished. A constant of 500 ms was added to the estimated RTO, leading to a more generous timeout estimation and

---

[16] One transport block set in the physical layer is encoded in a single block. Thus at the RLC receiver the granularity of the blocks is the same, a single transport block set.

reducing the effect of spurious retransmissions described above. The modified RTO calculation ensures that the congestion window can reach its upper limit. This guarantees full use of the available bandwidth.

Throughput performance of the 2-Mbit/s bearer service with the modified TCP Reno implementation is shown in Fig. 5.57. The observed max-

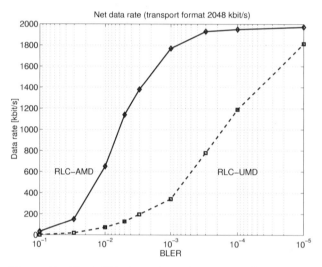

Fig. 5.57. Net TCP throughput for **2-Mbit/s** radio bearer. Both RLC acknowledged and unacknowledged mode (RLC-AMD and RLC-UMD) are plotted

imum throughput for RLC-AMD is 1974 kbit/s and for RLC-UMD about 1982 kbit/s (96.5% and 96.8% of raw data rate). The steepness of RLC-AMD and RLC-UMD curves are much more gentle than for all other transport services. To achieve 90% of the maximum throughput a BLER of 8×10^{-4} is required for RLC-AMD. This is more than one decade below the required BLER for other bearer services. The probability for multiple segment losses in a single transmission window is increased significantly compared to the other bearer services. The reason is again the much larger number of packets in a transmission window, firstly because of the large number of packets in a transport set, secondly because of the window-scaling mechanism which also double the transmission window. Thus, the probability for violation of TCP's 'Fast Retransmit' algorithm assumption grows, leading to lower throughput performance.

Influence of TCP/IP header compression. For the 64-kbit/s radio bearer, simulations with activated TCP/IP header compression have been performed. TCP/IP header compression is done in the PDCP layer. A comparison between TCP net throughput with and without activated header

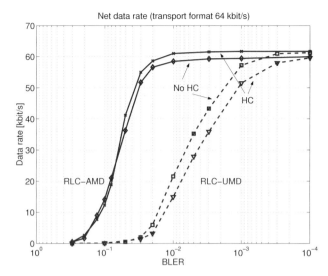

Fig. 5.58. Comparison between net TCP throughput of the **64-kbit/s** radio bearer with and without activated **TCP/IP header compression** (HC). Both RLC modes (RLC-AMD and RLC-UMD) are shown

compression is plotted in Fig. 5.58. Both RLC modes (RLC-AMD and RLC-UMD) have been evaluated.

Figure 5.58 gives us very interesting results. The curves without header compression (HC) show almost the same performance in terms of required BLER as the 144-kbit/s radio bearer. All conclusions drawn for the 144-kbit/s radio bearer are also valid for the 64-kbit/s radio bearer. The maximum throughputs with and without HC differ because of the slightly lower overhead. It is 61.71 kbit/s for RLC-AMD with HC compared to 60.15 kbit/s for RLC-AMD without HC (96% compared to 94% of raw data rate). Accordingly, 62.45 kbit/s and 61.34 kbit/s maximum throughput for RLC-UMD are observed, which corresponds to 97.6% and 96% of raw data rate[17].

Activation of TCP/IP header compression has a *contradictory* influence on RLC-AMD and RLC-UMD. The performance of RLC-AMD *increases*, in general, while performance of RLC-UMD *decreases*.

TCP throughput for RLC-AMD with HC exceeds the one without HC over a large BLER area. However, for BLER worse than 8×10^{-2}, RLC-AMD with HC has lower throughput than RLC-AMD without HC. At this point, the relatively high BLER causes TCP to send timeout-triggered retransmissions. As the HC sender entity receives this retransmissions, the HC sender entity sends an uncompressed packet because the retransmitted packet

---

[17] The maximum values for RLC-UMD are not plotted in Fig. 5.58 because scaling ends at a maximum BLER of BLER $= 10^{-4}$. There exists a crossover of the two curves for RLC-UMD for BLER $< 10^{-4}$.

is *out-of-sequence* of all other packets sent before. Thus throughput does not gain from compressed packet sizes. Also, if compressed packets get lost, the HC transmitter entity does not know this immediately. Instead the HC receiver entity needs to inform the HC sender by sending an update packet back to the HC sender. Of course, this will take some time. In this time span, the HC sender will produce compressed packets which can *not* be decompressed by the HC receiver, because a packet was missing[18]. Thus performance gets worse with activated HC when BLER exceeds a specific value.

The same reason as explained for RLC-AMD causes the poor performance of RLC-UMD with HC. Throughput is always lower when HC is activated. If many blocks are in error, frequent unsynchronized states between HC compressor (PDCP sender) and HC decompresser (PDCP receiver) occur. Thus many received packets are incorrectly decompressed, which leads to TCP packets with wrong check sum. These packets have to be discarded, they waste bandwidth.

As a conclusion, TCP/IP header compression should be activated *only* if the underlying transmission medium is reliable. This means in UMTS, if the RLC-AMD mode is used. For RLC-UMD, HC should be deactivated.

FTP and WWW traffic examples. In this section we will present two example scenarios for FTP and WWW traffic.

The first example consists of a simultaneous FTP *up-* and *download*. Two independent FTP clients and servers, one pair for FTP upload and one pair for FTP download, are active. Every FTP client-server pair uses independent TCP entities. The ability of TCP to piggyback data on acknowledge packets is therefore *not* used. The instantaneous throughput for both connections is shown in Fig. 5.59. RLC-AMD mode together with a loss-free channel (BLER=0) had been used. The FTP download connection starts at $t = 0$ and maintains throughput the entire simulation. FTP upload starts at $t = 200\,\mathrm{s}$ and finishes at $t \approx 500\,\mathrm{s}$.

Mean TCP throughput for exclusive FTP download is 368.3 kbit/s. When FTP upload is active too, mean throughput reduces to about 275 kbit/s, which is only 75% of the maximum TCP throughput of a single connection. Thus 25% of the bandwidth is occupied by the acknowledge traffic of the other connection. This is a clear indication that the acknowledge traffic must *not* be underestimated.

The glitches at the starting and endpoint of the FTP upload occur due to timeout-triggered retransmissions. These are caused by the sudden changes of the bandwidth in the path which also implies a change of the packet delay.

---

[18] TCP/IP header compression is a delta compression method. Only the differences to the previous packets are transmitted. Thus, if a packet gets lost, information is missing. Packets arriving after a packet loss cannot be decompressed properly. But the receiver does not recognize this immediately, because the real information is not known to it. It has to wait for an uncompressed packet to be synchronized again.

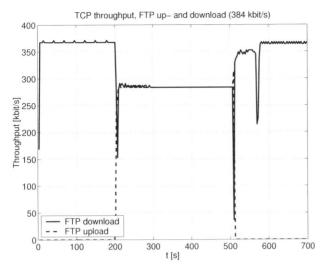

Fig. 5.59. Partly simultaneous FTP up- and download on a single 384-kbit/s radio bearer. FTP upload starts at $t = 200$ s and ends at $t \approx 500$ s. FTP download is active the entire time, loss-free channel is assumed

Thus, TCP needs some packets to detect the new available bandwidth and adjust the RTT and RTO estimation to the new situation.

As a second example we will present WWW traffic evaluation. The parameters of the WWW traffic model are given in Table 5.13. A *single* HTTP client on a mobile station requests files from a HTTP server. The channel BLER was set to BLER $= 1 \times 10^{-2}$ and the external packet data network IP packet loss to BLER$_{\text{Internet}} = 2 \times 10^{-3}$. As there is only a single user, only a single file at one time instant is transmitted.

Figure 5.60 shows the TCP throughput per transmitted file versus the file size. A significant percentage of the requested files is limited at an 'upper bound' of the throughput. This 'upper bound' is given by the maximum TCP throughput for large files. For file sizes below 100 kbyte, however, throughput does not reach the theoretical optimum as for a long maintaining file transfer. If the time between two HTTP requests is long enough for TCP to fall back into the idle mode, the *slow start* mechanism is activated. Thus the TCP transmission window has to grow up to the maximum for every file again which reduces average throughput. This effect is more pronounced at small file sizes and does not have a significant effect for large files.

As a large amount of files reach this 'upper bound' in throughput, the system is not fully loaded. This is, there are idle times between transmissions of files. If requests occur when there is already a file transmitted, throughput decreases. This is indicated by the sample points in the area below the 'upper throughput bound'.

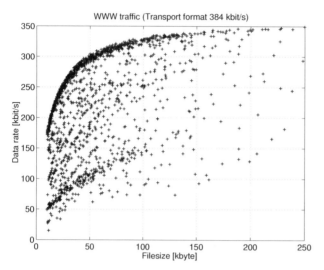

Fig. 5.60. Simulated throughput per file over the file size for a single user. A WWW traffic model according to Table 5.13 was used for file generation. The 384-kbit/s radio bearer and RLC-AMD were used, BLER was set to $BLER = 1 \times 10^{-2}$ and external packet data network loss was $BLER_{Internet} = 2 \times 10^{-3}$

In Fig. 5.61 additionally the evaluation for *the same* client is shown, but a second, similar behaved client is active in the background on a different TCP connection. This new evaluation is labeled '2 clients'. For comparison also the result from Fig. 5.60 is plotted (labeled '1 client').

No significant difference can be observed between the two cases. However, looking at the cumulative distribution function (CDF) gives more insight. The CDF of the throughput values for all file sizes is drawn in Fig. 5.62. For a large throughput range from 60 kbit/s up to 300 kbit/s the two active clients have more *lower* throughput values than a single client. For example, at a level of $CDF(x) \approx 30\%$ about 25 kbit in throughput difference is observed. Correspondingly, at very low throughput levels (< 50 kbit/s) probability for this low throughput values is higher for 2 clients than for a single client.

But the differences are nevertheless relatively small. Thus the 384-kbit/s radio bearer is *by far not* fully loaded with two WWW clients according to the model specified in Table 5.13. Of course, using a completely different WWW traffic model may yield to different results.

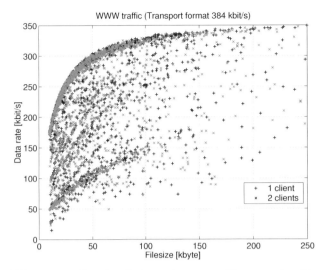

Fig. 5.61. Simulated throughput per file over the file size for one user, when a second, similar behaved user is active and produces background traffic (label '2 clients'). The evaluation from Fig. 5.60 is plotted for comparison (label '1 client'). A WWW traffic model according to Table 5.13 was used for file generation, simulation parameters are equal to those used in Fig. 5.60

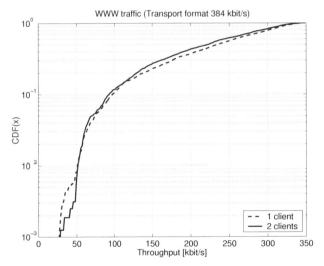

Fig. 5.62. CDF of throughput per file for a single user (label '1 client') and for a user, when a second, similar behaved user is active and produces background traffic (label '2 clients'). A WWW traffic model according to Table 5.13 was used for file generation (exponential file size distribution)

5.6.3 Summary and Conclusions

We presented simulations of TCP/IP data transmission over UMTS radio bearers [92]. FTP (bulk data transfer) and a sample WWW scenario were investigated.

Throughput performance. A summary of throughput performance of the investigated radio bearers is shown in Table 5.14. Both RLC-AMD and RLC-UMD results are listed. The 2-Mbit/s radio bearer is excepted here.

Table 5.14. Comparison of UMTS bearer service performance for RLC-AMD and RLC-UMD. The 64-kbit/s radio bearer was simulated, but not presented in the text.

Transport format	RLC-AMD		
		Throughput	BLER at 90%
	max.	relative to raw	of max. throughput
(kbit/s)	(kbit/s)		
12.2	10.45	86%	4×10^{-2}
64	60.15	94%	2.5×10^{-2}
64 (incl. HC)	61.71	96%	3×10^{-2}
144	136.8	95%	1.8×10^{-2}
384	368.3	96%	1.2×10^{-2}
2048	1974	96.4%	8×10^{-4}
	RLC-UMD		
12.2	11.1	91%	1×10^{-3}
64	61.35	96%	2×10^{-3}
64 (incl. HC)	62.45	97.6%	4×10^{-4}
144	139.03	96.5%	2×10^{-3}
384	371.3	96.7%	1.5×10^{-3}
2048	1982	96.8%	5×10^{-5}

In general, the following conclusions can be drawn:

- RLC-AMD needs a BLER about BLER $= 1.2 \times 10^{-2}$–4×10^{-2} to achieve 90% of the maximum throughput.
- RLC-UMD needs about one order of magnitude lower BLER than RLC-AMD (BLER $= 1 \times 10^{-3}$–4×10^{-4}) to achieve 90% of the maximum throughput.
- Relative throughput is higher for radio bearers with larger transport block size (higher bit rate corresponds with larger transport block size), because the protocol overhead is smaller. Radio bearer 12.2 kbit/s has a comparably lower transport block size than in GSM RLP. Overhead is very large which explains the low relative throughput.

- RLC-AMD and RLC-UMD have different overhead ratios. This is *different* from GSM-GPRS, where the overhead is the same in RLC acknowledged and unacknowledged mode.
- When considering realistic BLER from BLER $\approx 1 \times 10^{-3}$ to 1×10^{-2} RLC-UMD is *not* suitable for reliable data transfer. It should be used for real-time traffic only, where block errors can be handled.

When comparing the results for different radio bearers one has to take into account the used error model. We only investigated a block error model. BLER at a specific SNIR on the channel might be completely different for different radio bearers. It will be larger for high-rate bearers because of their large transport block size. Thus relative performance may vary when considering realistic BLER together with system simulations. Especially in a CDMA system like UMTS link performance varies heavily with the number of users. Thus combined study of link and system simulations is necessary to achieve performance results under realistic scenarios.

2-Mbit/s radio bearer. The 2-Mbit/s radio bearer has to be considered as a *long, fat network*, because the bandwidth-delay product exceeds the maximum size possible for standard TCP parameters under realistic conditions.

A second, more severe problem is introduced by the transport format. One transport block set consists of 40 transport blocks with 4096 bits. Thus more than 20 kbytes of data is transmitted and received at 'a single' time instant. TCP is self-clocking and relies on regularly arriving acknowledgment packets. Now the acknowledgment packets arrive in large bursts. This induces serious errors in the RTT estimation and thus also the RTO estimation. False RTO estimation leads to spurious timeouts which degrades performance. The above statements are valid for TCP Reno. The new TCP 'Vegas' version may perform better, however, this is not quite clear and needs further investigation.

To achieve maximum throughout the following points have to be considered:

- The *Window scaling* option of TCP is *mandatory*. This options increases the maximum possible number of unacknowledged data packets in transit.
- TCP Reno's RTT (and RTO) estimation needs to be modified. In [92] a brute-force approach was tried. The RTT and RTO estimation was not altered, but only a constant of 500 ms was added to the estimated RTO. This prevents the spurious timeouts, but also decreases performance at very poor BLER. Clearly, this is no optimum solution. Again TCP Vegas might perform better but this has not been verified.

We conclude:
Window scaling is mandatory for the 2-Mbit/s radio bearer. Usage of the 2-Mbit/s bearer service with standard TCP 'Reno' will not give satisfactory performance. An upgrade to TCP Vegas or a specially designed TCP version will be necessary.

Influence of TCP/IP header compression. The influence of TCP/IP header compression on the net performance has been investigated for the 64-kbit/s radio bearer (Fig. 5.58). A summary of the results is also listed in Table 5.14.

- Maximum throughput *increases* due to the smaller protocol overhead because of the compressed header.
- Activated TCP/IP header compression has a contradictory influence on RLC-AMD and RLC-UMD compared to data transmission without any header compression at all. When activating TCP/IP header compression,
  – RLC-AMD performance *increases*,
  – RLC-UMD performance *decreases*,
  for a specific BLER.

> **We conclude**:
> TCP/IP header compression is advantageous when operating with a RLC-AMD bearer and should be used in conjunction with RLC-AMD.
>
> When RLC-UMD is used, TCP/IP header compression should be *deactivated*.

FTP and HTTP traffic. Two sample scenarios with simultaneous FTP up- and download on a single radio bearer, and WWW traffic investigation for one and two clients have been presented. In both studies the 384-kbit/s radio bearer and RLC-AMD mode was used.

When doing simultaneous FTP up- and download on a single radio bearer the resources are shared between the data packets of one connection and the acknowledgment packets of the second connection. The acknowledgment packets decreases the throughput by about 25% compared to the maximum throughput when a bearer is exclusively used by a single connection.

A simple WWW traffic model (parameters are listed in Table 5.13) was used to obtain throughput values for sample WWW traffic files. For file sizes below 100 kbyte TCP's *slow start* mechanisms dominates the data transfer and reduces the mean throughput. The throughput distribution did not change significantly when a second, similarly behaved client operated in the background. Thus the system was not fully loaded. However, a slight shift of about 25 kbit/s towards lower throughput was observed in the distribution function of the throughput values.

6. Summary and Conclusions

Mobile Internet access and mobile e-commerce are the major drivers to develop packet-switched mobile radio networks like the *General Packet Radio Service* (GPRS) and the packet-switched services in the 3rd generation mobile radio system *Universal Mobile Telecommunication System* (UMTS). However, to familiarize the costumers to mobile Internet access and also to achieve revenue as soon as possible, network operators push the mobile Internet access also on circuit-switched data services in the 2nd generation mobile radio networks. The most successful 2nd generation digital mobile radio system is the *Global System for Mobile communications* (GSM).

As pointed out already in the introduction (Chap. 1) basically two methods exist for wireless Internet access via cellular phones:

- Usage of the mobile phone as a 'wireless modem' to connect a standard computer with standard networking software to the Internet
- Usage of dedicated software (e.g. *Wireless Application Protocol*) on the mobile phone to access the Internet in a predefined manner.

The first method shines due to its standard network software which enables utilization of all standard (networking) applications currently used on every computer. However, the network protocols IP and especially TCP have been developed for nearly error-free links. TCP assumes that 99% of the packet losses are caused by network congestion rather than link failure. Thus their performance over an erroneous radio link is not optimal.

In Chap. 2 we investigated the second approach, WAP, and its advantages and deficiencies. Currently it provides only rather limited access to the Internet, so we stressed, in this book, the first approach and the impact of the unreliable radio link on the performance of the Internet protocols, especially on TCP.

Therefore we simulated the performance of a TCP/IP data transmission over GSM circuit-switched links, and GSM-GPRS and UMTS packet-switched links. To assess the GSM and GPRS performance we developed a bit and block error model for the GSM physical layer. In either standard, we implemented the complete protocol stack from physical layer up to the transport layer, the TCP protocol. The results and conclusions which we draw from this investigation are summarized in the following.

6.1 Wireless Application Protocol

The Wireless Application Protocol provides a protocol stack from the transport layer (layer 4) up to the application layer. It is a set of protocols, applications environments and content formats which is built on top of current and future bearer services. The most important bearer services are GSM *Circuit-Switched Data* (CSD), GPRS and, in the future, UMTS bearers. The lowest-level WAP protocol (*Wireless Datagram Protocol*, WDP) adapts the bearer to a common higher layer interface. All higher-layer protocols operate independently from the currently used bearer.

WAP is based on existing protocols, applications and content formats. But they are modified to meet the demands for lightweight mobile applications and the special demands for operating over the unreliable radio link.

The WAP model of operation is *transaction* based, similar to the WWW concept of a client-server model. The client requests information from a server, but the request is processed at first at an intermediate gateway where the request is decoded and sent to the actually addressed server. The server sends the requested information back via the gateway to the client. Request and response are sent encoded over the radio interface for efficient resource utilization.

Basic security features are provided by a special layer based on standard cryptographic algorithms and protocols.

A special description language, *Wireless Markup Language* (WML), is used to present the content in a way similar to HTML used in ordinary Internet.

WAP is optimized for short transactions with a small amount of exchanged data. The transaction layer protocol *Wireless Transaction Protocol* (WTP) provides reliable and unreliable transactions. However, only very rudimentary elements for *flow control* of data are implemented. Thus WTP is *not* suited for large-scale data transmission. This may be a problem if several WAP clients are served on the same base station and request significant amount of data at the same time. If, therefore, a congestion in a network node occurs (possibly at the base station or the *Serving GPRS Support Node* in GPRS) no means for efficient congestion control are defined. WTP will *not* detect this as a problem.

Thus a well-dimensioned WAP backbone network is mandatory. This could also be a bottleneck in WAP for the usage in future high-performance mobile networks like UMTS.

Also, the limited presentation capabilities of WML could be an obstacle for the usage of WAP in the future, when the data capabilities have emerged to provide full-featured Internet access.

6.2 GSM and GSM-GPRS Radio Link

We implemented the entire physical layer (including channel coding, interleaving, burst formatting, modulation and demodulation) in a simulation model to assess GSM and GPRS performance under realistic radio environments. As a channel model we used the COST 207 wideband channel model which consists of four typical environments, 'Typical Urban' (TU), 'Bad Urban' (BU), 'Hilly Terrain' (HT) and 'Rural Area' (RT). We evaluated the performance in terms of BER by means of link level simulations. Results are available for the GSM-CSD traffic channels TCH/F9.6 and TCH/F14.4 as well as for all four GPRS coding schemes CS-1 to CS-4.

TCH/F9.6 has the best performance and yields a BER $= 10^{-5}$ at a $C/I = 10\,\mathrm{dB}$ for a mobile station speed of $v = 10\,\mathrm{km/h}$ and the TU channel model. This is almost one order of magnitude better than TCH/F14.4. The best GPRS coding scheme, CS-1, achieves about the same performance as TCH/F14.4. Although CS-1 has the same code performance as TCH/F9.6 it suffers from the worse interleaving as do all GPRS coding schemes. All other GPRS coding schemes, CS-2 to CS-4 are worse than CS-1. The range is about 1.5 ($v = 250\,\mathrm{km/h}$) up to 2 ($v = 10\,\mathrm{km/h}$) orders of magnitude in BER compared to CS-1. GPRS CS-4 coding scheme is virtually uncoded and yields very poor performance, a minimal BER of about 2×10^{-3} and a block error ratio (BLER) of only BLER $\approx 7 \times 10^{-1}$ was obtained. BER performance evaluated with the bad urban (BU) channel model is about one order of magnitude below that for TU. We used the TU model for all subsequent evaluations, especially for TCP.

We modeled the entire link layer with a 2-state Hidden Markov Model (HMM). We showed that the interleaver in GSM is powerful enough to combat all burst errors. Thus a link level model only needs to be able to provide independent identically distributed (i.i.d.) error patterns. The dependency of BER (BLER) on C/I can be expressed with a simple 'low-pass filter' function. The parameters of the function are obtained by curve fitting. This leads to a simple link level model which enables extensive and time-saving evaluation of higher-level protocols.

6.3 GSM-GPRS vs. UMTS Packet-Switched Data

GPRS relies on the GSM access network (FD/TDMA) and reuses its radio resources. Circuit- and packet-switched data use the same timeslot structure of 8 timeslots per TDMA frame. Upgrading from GSM to GPRS base stations requires, in the best conceivable case, only a software upgrade. GPRS provides data rates up to nominally 172 kbit/s with no channel coding and when utilizing all 8 timeslots per frame.

The UMTS access network is based on Direct Sequence (DS) CDMA. Two different duplex modes exist, *Frequency Domain Duplex* (FDD) and *Time*

Domain Duplex (TDD). We investigated mainly the FDD mode. UMTS data rates are very flexible due to the CDMA technique and a flexible multiplexing of transport channels onto the physical channels.

The UMTS packet-switched data core network is essentially an evolved GPRS backbone network. The core network architecture for Release'99 of UMTS is almost identical to that of GPRS. Due to the different access networks, some changes in the locations of various functions have been made:

- In UMTS the logical link ends in the *UMTS Terrestrial Access Network* (UTRAN) in the *Radio Network Controller* (RNC). In GPRS the link spans onward to the SGSN.
  Thus a *3rd Generation Serving GPRS Support Node* (3G-SGSN) has no logical link management functions any longer.
- UMTS provides no *data* compression any longer in contrast to GPRS, where this task was handled by the *Subnetwork Dependent Convergence Protocol* (SNDCP).
- Header compression in UMTS is accomplished now entirely in the UTRAN (RNC). In GPRS this was also handled in the SGSN. This stems from the fact that the logical link already ends in the UTRAN (see Item 1).
- Ciphering of user data is handled by the SGSN in GPRS as well. In UMTS it is shifted to the RNC entity. Additionally the ciphering function is moved a layer down to the *Radio Link Control* (RLC) or even *Medium Access Control* (MAC) layer compared to the *Logical Link Control* (LLC) layer in GPRS.
- Radio resource management has been moved to UTRAN from the SGSN, where it is located in GPRS.
- The RLC *acknowledged mode* and *unacknowledged mode* have *equal* protocol overhead in GPRS. This has changed in UMTS, where the RLC-*Unacknowledged Mode Data* (UMD) has lower protocol overhead than RLC-*Acknowledged Mode Data* (AMD).

In general, UMTS has more functionality within the access network than in the core network. Many functions have moved from SGSN to RNC compared to GSM GPRS. Thus UTRAN is more decoupled from the core network than was the case in GSM GPRS.

This has advantages when connecting the UTRAN network to a future UMTS core network. The interface for packet-switched data between UTRAN and UMTS core network (*IuPS*) is a strict border between these parts.

6.4 TCP Performance in Wireless Environments

In Chap. 5 we assessed the performance of TCP when operating over wireless links. In a first step we provided an overview of the known TCP problems on an unreliable link and we gave an overview of solutions suggested in the literature. We computed the optimum block length for a go-back-N *Automatic Repeat Request* (ARQ) protocol (like TCP) for transparent mode operation. Finally, we simulated TCP/IP data transmission over GSM, GPRS and UMTS. We also performed measurements of data transmission over live GSM links.

6.4.1 Problems of TCP and Suggested Solutions

We classified the problems of TCP into four classes which we summarize in the following:

- The basic assumption of TCP that 99% of all packet losses are caused by congestion in the network is violated when operating over a highly unreliable medium like a radio link. Thus TCP's reaction to a packet loss is in most cases wrong. TCP assumes a congestion and decreases the amount of packets injected into the pipe by decreasing the *congestion window*. Thus throughput performance is decreased because the packet injection rate is decreased. In the ideal case, TCP should retransmit only the lost segment and inject the packets with the same rate as before the loss occurred.
- If a temporary disconnection of the link occurs, it depends whether this is *lossy* or *non-lossy*. For lossy disconnections (e.g. due to lossy handovers or link errors), TCP reacts as in Item 1. The congestion window is reduced and TCP retransmits the lost segments. The better solution would be to retransmit the lost segments *without* reducing the congestion window.
  For non-lossy disconnections (e.g. handovers) it depends on the time of the disconnections. If it is long enough so that TCP's retransmission timer expires, a retransmission is scheduled whether it is necessary or not. TCP reacts again as if a packet loss had occurred.
- If a lower layer reliable link level protocol is active in addition to TCP, the interactions between both protocols might have a negative influence on TCP's throughput performance.
  If the link level protocol does its own retransmissions the latencies of the packets may vary. This misleads TCP's *Round Trip Time* (RTT) estimation and may lead to spurious timeouts at the TCP sender.
  A different effect is that the variance of the RTT value gets very high which leads to a large retransmission timeout. Thus TCP's reaction time on *real* packet losses is extended significantly and may decrease overall performance.
- In asymmetric systems (downlink bandwidth is much higher than uplink bandwidth) the acknowledgment traffic on the uplink may be higher than

the available uplink bandwidth. We presented two examples for GPRS and UMTS-TDD where the required uplink bandwidth is close to or even higher than the available one.

Solutions for this problem are either the delayed acknowledgment mechanism (acknowledging only a fraction of the data segments), a special ack congestion control algorithm, or an acknowledge filtering method.

Many solutions to the problems reported above are available in the literature. A short summary of those which we presented in Sect. 5.2 is given in the following. Basically all solutions can be classified into three schemes:

- End-to-end proposals:
  End-to-end methods try to improve TCP's performance by improving TCP itself. Only the TCP sender and receiver are modified.

  Methods already included in TCP are the *fast retransmission* and *fast recovery* algorithms, which provide a good solution if only single packets are lost.

  An important method is also the *selective acknowledgment* (SACK) option for TCP. With this, only those segments that are really lost are retransmitted, and not all subsequent packets as well. However, a problem with TCP SACK is the interaction with TCP/IP header compression. Header compression performance is decreased because the TCP/IP header changes often when SACK is used, thus differential compression does not work well. Additionally the overhead is increased significantly.

  Some proposals make TCP aware of the mobility, e.g. when handovers occur, if the link is disconnected or if link quality is poor. Although the performance improvement is sometimes rather high, this includes major changes in the TCP protocol itself, which is not desirable.
- Split-connection proposals:
  In the split-connection approach, *separate* TCP connections between the MS and an intermediate host, and the intermediate host and the fixed host are established. At the connection to the MS an optimized TCP protocol can be used. However, in a straightforward implementation, the end-to-end semantic of the transport layer protocol is violated. Also, buffer overflow at the intermediate host may be a problem.

  Some proposals try to fix the semantic problem with methods like examining the data stream in both directions.

  But, it has been shown for an early representative of these class of solutions ('the Snoop protocol') that it does not work well in a busy GPRS system.
- Link-layer proposals:
  Link layer protocols provide a reliable connection at the data link layer. GSM, GPRS and also UMTS have their own link layer protocols over the radio interface. These are optimized, selective repeat ARQ protocols. The advantage is that the transport layer (TCP) needs not to be changed. The maximum throughput in simulations was 8 kbit/s. TCH/F14.4 needs about 5 dB more in C/I. TCH/F9.6 has a lower C/I threshold.

6.4.2 TCP Packet Length for Optimum Throughput

For transparent transmission modes (i.e. TCP is the only reliable transmission protocol, there is no active, reliable link layer protocol) an *optimum* TCP block length exists. With this block length, the *maximum throughput* is achieved. Two competing effects are responsible for the optimum block length: (1) When packet length increases, BLER increases, thus throughput *decreases*. (2) For increasing block length, the overhead due to constant header length decreases, thus throughput *increases*. Additionally the optimum block length depends on the Round Trip Time (RTT).

In general, block length should be larger for better link quality and higher header length. When the RTT increases, optimum block length decreases, but not to a huge extent.

We calculated the optimum block length for a go-back-N ARQ protocol (e.g. TCP) for three different link error characteristics.

- Binary symmetric channel with error-free feedback:
  The optimum block length for a standard TCP/IP header length of 40 bytes is about 100–1000 bytes for BER $\approx 10^{-4}$–10^{-6}.
- Markov error model with error-free feedback:
  We estimated 2-state Markov error model parameters from link level simulation error patterns. For a GSM CSD TCH/F9.6 data traffic channel and a C/I range from $C/I = 10$–20 dB the data block length should be selected to $n_{\mathrm{D}} \approx 20$–$200$ bytes.
- Markov error model on forward and feedback links:
  Both forward and feedback links were modeled with a 2-state Markov error model. For the GSM TCH/F9.6 traffic channel a block length from $n_{\mathrm{D}} \approx 100$–$1000$ bytes for $C/I = 10$–20 dB should be selected.

A summary of the results is given in Table 6.1.

Table 6.1. Optimum data block lengths in bytes for different error characteristics. The RTT was set to $RTT = 4$

Error characteristic forward channel	feedback channel	Optimum data block length (bytes) Header length: 5 bytes	40 bytes	Corresponding C/I (BER) range
Binary symmetric	ideal	40–300	100–1000	BER $\approx 10^{-4}$–10^{-6}
Markov	ideal	10–80	20–200	$C/I \approx 10$–20 dB
Markov	Markov	40–180	100–1000	$C/I \approx 10$–20 dB

6.4.3 GSM Circuit-Switched Data

To assess the performance of TCP/IP data transmission over GSM CSD traffic channels we performed simulations and measurements. The simulations were done in transparent mode for TCH/F9.6 and TCH/F14.4 with SLIP as link level protocol. In the measurements we utilized TCH/F9.6, TCH/F14.4, and the HSCSD traffic channels TCH/F19.2 and TCH/F28.8. We will summarize the most important results in the following:

- In *transparent* mode, TCH/F9.6 achieves 70% of maximum **throughput** for $C/I \geq 15$ dB. The maximum throughput in simulations was 8 kbit/s. TCH/F14.4 needs about 5 dB more in C/I.
- Measurements in transparent mode yielded only a maximum of 6.6 kbit/s (86%) for TCH/F9.6.
- *Non-transparent* mode measurements achieved 8.3 kbit/s (86%) with SLIP and about 8.6 kbit/s (89%) with CSLIP as link layer protocol in TCH/F9.6.
- TCH/F14.4 and TCH/F28.8 yielded 80% to 85% of the theoretical throughput in stationary measurements. This is about 5% below the the relative throughput achieved with TCH/F9.6. The worse channel coding for TCH/F14.4 and TCH/F28.8 is responsible for this degradation.
- Absolute throughput is almost always higher for TCH/F14.4 than for TCH/F9.6. As the resource utilization is the same for both traffic channels (a single timeslot), we recommend to use TCH/F14.4 to maximize throughput.
- **TCP/IP header compression** should be activated, when *non-transparent* mode is used. See also conclusions for UMTS/64-kbit/s bearer with and without header compression.
- The **optimum block length** in *transparent* mode, theoretically predicted in Sect. 5.3 was verified by both simulations and measurements (see Table 6.1). To achieve optimum throughput in applications that need transparent mode, the optimum TCP block length should be used. Such an application could be, for example, a real-time application, which uses usually UDP but also has the need to transmit reliable information with TCP. The optimum block length for TCH/F9.6 is 500 to 800 bytes. The worse the channel quality, the smaller the block length should be set.
- Although PPP protocol overhead is slightly higher than CSLIP overhead, the influence is only marginal. Thus we recommend to use PPP due to its easy and more sophisticated link establishment and parameter negotiation.
- We performed **RTT measurements** with 92 bytes ICMP packets for both TCH/F9.6 and TCH/F28.8. About 90% of all RLP packets experienced *no* retransmission which results in a RTT of about 900 ms. For TCH/F28.8 the mean RTT for the majority of the packets was 880 ms. The difference is about the difference in pure transmission speed at the radio interface.
  We conclude that the main part of the delay is introduced in the core network of the mobile network, the IWF and the external packet data

network. The delay on the radio interface is only a minor part. Thus it is important to optimize the backbone network of the mobile network as well as to improve transmission quality on the radio link.

- In **wide area measurements** the throughput decreased by up to 37% compared to the stationary measurement results. The only exception was the 'rural area, parking place' scenario where a degradation of only 13% was observed.
- Higher-layer protocols involved during connection setup phase (PAP and PPP) introduce serious problems in wide area measurements. Although the GSM link establishment worked very reliably, the subsequent network log on, authentification and PPP protocol negotiation phases had frequent errors especially when using the worse channel coding schemes (TCH/F14.4 and TCH/F28.8). These protocols seem to have problems with frequent disconnections and long delays.

This needs further investigation to provide well-founded results.

6.4.4 GSM Packet-Switched Data (GPRS)

We investigated the TCP/IP performance of a GPRS system for the single and multi user case. All four coding schemes (CS-1, CS-2, CS-3 and CS-4) were assessed.

The most important results regarding the **single-user** case are:

- CS-1, CS-2 and CS-3 achieve about 70% of the maximum theoretical gross rate for RLC *acknowledged* mode. CS-4 performance is inferior. The threshold C/I required to achieve more than 90% of the maximum throughput is for CS-1 at $C/I_{\text{thres}} = 10\text{--}15\,\text{dB}$, whereas for CS-2 and CS-3 it is shifted to $C/I_{\text{thres}} = 20\,\text{dB}$.
- RLC acknowledged mode protocol might not be suited very well for transmission over the maximum bandwidth of 8 timeslots. The window size is fixed at $k = 64$ which seems to be too small if operating over 8 PDCHs.
- In RLC *unacknowledged* mode only CS-1 performs well and again achieves about 70% of the maximum theoretical gross throughput. C/I threshold to achieve 90% of the maximum throughput is shifted about $10\,\text{dB}$ to higher C/I for unacknowledged mode compared to acknowledged mode.
- In the particular environment that we investigated ('Typical Urban', $v = 10\,\text{km/h}$) *only* RLC acknowledged mode is suitable for *reliable* data transmission with TCP.
- Packet delay is, in general, lower for RLC unacknowledged mode. It might give sufficient performance for applications that can cope with relatively high error rates, but need predictable packet delays (real-time applications, voice and video).
- The scaling of the throughput, when multiple timeslots are used compared to a single timeslot, is linear. However, for 8 timeslots performance is

slightly worse, which might be due to the fixed RLC window size (see second item above in this list).

- Measurements in the literature agreed very well with the results we obtained by means of simulation for coding scheme CS-2 and 2 timeslots in downlink. Thus similar results can be expected for other timeslot combinations and different coding schemes.

For assessing the system performance under the load of **multiple users** with WWW and also mixed WWW-circuit-switched traffic we assumed a typical WWW traffic model and empirical circuit-switched traffic assumptions. The absolute values depend on the traffic model parameters, but the qualitative behavior of the system is independent of them. The system performance is the cumulative throughput of all users either on RLC or on TCP level.

Summarizing the results:

- When the number of WWW users in a system without circuit-switched traffic is below a specific *critical value*, the system throughput performance scales linearly with the number of users.
- If the number of users exceeds the critical value, RLC layer throughput is stable at the maximum throughput. But *cumulative* TCP system throughput of all users *decreases* when the number of users increases. Thus the system *must not* be overloaded if TCP traffic is transmitted.
- The critical number of users is 5 to 20 WWW users (without circuit-switched traffic) for 1 to 4 PDCHs.
- The total system performance is *independent* of the multislot capability of the mobile stations. The individual user will benefit from a higher multislot class at the cost of another user, but system performance is not affected.
- For *mixed packet- and circuit-switched* traffic the critical number of WWW users is roughly *one half* compared to the isolated packet-switched case. This result is obtained for empirical parameters for the circuit-switched traffic observed in a real GSM network.

6.4.5 UMTS Packet-Switched Data

We presented simulation results for TCP/IP data transmission over UMTS packet-switched data [92]. Several transport formats (radio bearers) have been investigated. The entire link level including channel coding was modeled with a block error model. Thus only results in terms of BLER are available.

The following main results are important:

- When using RLC acknowledged mode data (RLC-AMD) the required BLER ranges from BLER $= 1.2 \times 10^{-2}$–4×10^{-2} to achieve 90% of the maximum throughput.
- RLC unacknowledged mode (RLC-UMD) needs about one order of magnitude lower BLER than RLC-AMD.

- Relative overhead depends on the transport block size which is different for all radio bearers. Low-rate radio bearers have a larger overhead and thus relatively lower throughput.
- The 2-Mbit/s radio bearer performance is significantly worse compared to the other radio bearers. The large transport block size of more than 20-kbyte introduces serious problems for TCP and its RTT and RTO estimation. Current standard version of TCP (TCP Reno) does *not* perform well over this bearer. The RTT (RTO) estimation needs to be modified. An upgrade to a specially designed TCP version or future TCP version (TCP Vegas) seems to be necessary.

  The large bandwidth turns this bearer also into a so-called *long-fat* network. Thus *TCP Window scaling* is mandatory.

  A **possible solution** to this problem might be to apply *traffic shaping* on the PDCP layer. The individual TCP/IP packets included in one transport block set should be split and an artificial gap between them should be introduced. Thus the individual TCP packets do *not* arrive in bursts any longer but rather in a constant, regularly spaced manner, which does not confuse TCP's RTT estimation any longer.
- TCP/IP header compression increases performance over RLC-AMD *only*. If header compression is used over an unreliable media as RLC-UMD, throughput performance is *decreased*.

  Thus TCP/IP header compression should be used only if the underlying transmission media (all protocol stacks below TCP) is already *reliable*.

6.5 General Comments

Comparison Between GPRS and UMTS Performance

Due to the entirely different link levels in GPRS and UMTS, it is hardly possible to compare the performance of packet-switched data between GPRS and UMTS. To do such a comparison, we abstracted the link layer completely and used the BLER as common basis. The comparison of the *required* BLER to achieve 90% of the maximum throughput is shown in Fig. 6.1. RLC *acknowledged* mode in both GPRS and UMTS is assumed.

The required BLER for GPRS ranges from about $1 \times 10^{-2} \leq \text{BLER} \leq 1 \times 10^{-1}$, for UMTS it is $8 \times 10^{-3} \leq \text{BLER} \leq 7 \times 10^{-2}$. In other words, a BLER below 10% is required to have acceptable throughput performance.

One exception is the 2-Mbit/s radio bearer. It needs a BLER of $\approx 1 \times 10^{-3}$ to achieve 90% throughput performance. This is caused by the transport block problem already pointed out in the previous section.

This comparison assesses the *protocol performance* of the entire protocol stack down to the physical layer. The physical layer is included only with the influence of the block format which is delivered from/to the MAC layer to/from the physical layer. All other details of the physical layer are not

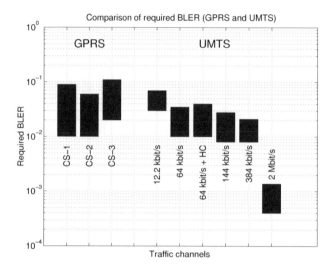

Fig. 6.1. Comparison between the required BLER of GPRS coding schemes and UMTS traffic channels to achieve 90% of the maximum TCP net throughput for RLC acknowledged mode

included. Considering this, it is not very surprising that GPRS and UMTS perform almost equally in terms of BLER. The radio link control protocols of both systems are similar but not identical.

The relative difference of the performance of UMTS radio bearers comes indeed from the influence of the physical layer due to different transport block sizes. Thus it makes sense to assess the protocol performance under realistic environments including the peculiarities of the physical layer. Only assessing the protocols for themselves does not reveal the full truth.

External Packet Data Network, Internet

When a user is downloading data from the Internet over a mobile radio link, normally the radio link is considered as bottleneck and as the weakest link. However, when we take a look at the current packet loss rates experienced in the Internet this might not necessarily be true.

In Fig. 6.2 the TCP/IP performance of the UMTS 384-kbit/s radio bearer under the assumption of a loss-free Internet and Internet block error rates, $BLER_I$, of $BLER_I = 2 \times 10^{-3}, 1 \times 10^{-2}, 2 \times 10^{-2}$ is plotted.

An Internet $BLER_I$ of 2×10^{-3} does not influence the performance significantly. However, on going to higher $BLER_I$, the overall throughput is dramatically reduced. Now the *radio interface is not* the limiting link, but rather the external Internet introduces too many errors.

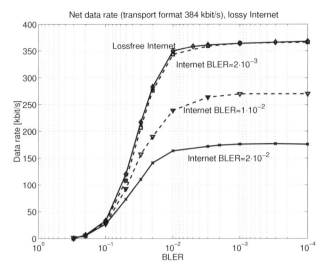

Fig. 6.2. Net throughput of TCP/IP data transmission over the UMTS 384-kbit/s radio bearer service for varied block error rates in the radio link. The Internet packet loss rate is the parameter. RLC acknowledged mode was used

Actual measurements in the global Internet yield a mean packet loss rate of about 2% to 4%[1]. Thus the external packet data network, in many cases the Internet, must not be underestimated when considering the end-to-end performance.

High-Performance Core Network

Introduction of high-rate radio bearers in GPRS and more severe in UMTS, and also the switching to packet-switched networks and services, will put heavy pressure on the core network. Many users have to be served and particular quality of service requirements must be met.

These demands can only be served if the core network undergoes the according upgrading as well. High-performance core networks and gateways to other networks, especially the Internet, will be needed to meet the future demands for quality of service for many users.

As we pointed out already in the previous subsection (Internet packet loss), core network packet loss due to wrong-dimensioned core network links may bring severe performance degradation even if the radio interface works reliably.

---

[1] See http://www.internettrafficreport.com.

A. List of Acronyms

3G-GGSN	3rd Generation GGSN
3G-SGSN	3rd Generation SGSN
AAL	ATM Adaptation Layer
ABM	Asynchronous Balanced Mode
AMD	Acknowledged Mode Data
AN	Access Network
ARQ	Automatic Repeat Request
ATM	Asynchronous Transfer Mode
AWGN	Additive White Gaussian Noise
BER	Bit Error Ratio
BLER	Block Error Ratio
BMC	Broad- and Multi-Cast
BSSGP	BSS GPRS Application Protocol
CDF	Cumulative Distribution Function
CN	Core Network
COST	European Cooperation in the Field of Scientific and Technical Research
CS-1	GPRS Channel Coding Scheme 1
CS-2	GPRS Channel Coding Scheme 2
CS-3	GPRS Channel Coding Scheme 3
CS-4	GPRS Channel Coding Scheme 4
CSD	Circuit-Switched Data
DCCH	Dedicated Control Channel
DCH	Dedicated Channel
DTCH	Dedicated Traffic Channel
EDGE	Enhanced Data Rates for GSM Evolution
ETSI	European Telecommunications Standard Institute
FACCH	Fast Associated Control Channel

FACH	Forward Access Channel
FDMA	Frequency Domain Multiple Access
FEC	Forward Error Correction
FER	Frame Error Ratio
FM	Frequency Modulation
GGSN	Gateway GPRS Support Node
GPRS	General Packet Radio Service
GSM	Global System for Communications
GTP	GPRS Tunneling Protocol
GTP-U	GPRS Tunneling Protocol, User Plane
HC	Header Compression
HDLC	High Level Data Link Control
HTML	Hypertext Markup Language
HTTP	Hypertext Transfer Protocol
ICMP	Internet Control Message Protocol
IETF	Internet Engineering Task Force
IP	Internet Protocol
IPCP	IP Control Protocol (NCP for IP in PPP)
IPX	Internetwork Packet Exchange (network layer protocol in Novell Netware operating system)
IPXCP	IPX Control Protocol (NCP for IPX in PPP)
IPv4	Internet Protocol, Version 4
IPv6	Internet Protocol, Version 6
IS-136	Interim Standard 136
ISDN	Integrated Services Data Network
ISI	Inter-symbol Interference
ISO	International Organization for Standardization
ISP	Internet Service Provider
ITU	International Telecommunication Union
LFN	Long Fat Network
LLC	Logical Link Control
MAC	Medium Access Control
MAC-b	Medium Access Control, for broadcast channels
MAC-c/sh	Medium Access Control, for common and shared channels
MAC-d	Medium Access Control, for dedicated channels
MLSE	Maximum Likely Sequence Estimator
MS	Mobile Station
MSC	Mobile-Services Switching Center

MT	Mobile Termination
MTU	Maximum Transmission Unit
NCP	Network Control Protocol (Sub-Protocol of PPP)
NRT	Non-Real Time
NTP	Network Time Protocol
OSI	Open Systems Interconnection
OSP	Octet Stream Protocol
OTA	Over-the-Air
PAP	Push Access Protocol (WAP push service)
PDCH	Packet Data Channel
PDCP	Packet Data Convergence Protocol
PDN	Public Data Network
PDP	Packet Data Protocol (e.g. IP)
PDP	Power Delay Profile
PDU	Protocol Data Unit
PID	Packet Identifier
PPG	Push Proxy Gateway (WAP push service)
PPP	Point to Point Protocol
PRACH	Packet Random Access Channel
PSTN	Public Switched Telephone Network
PTM	Point to Multi-point
PTP	Point to Point
QoS	Quality of Service
RA	Rate Adaptation
RAB	Radio Access Bearer
RACH	Random Access Channel
RLC	Radio Link Control
RLP	Radio Link Protocol
RNC	Radio Network Controller
RNS	Radio Network Subsystem
RT	Real Time
SAP	Service Access Point
SDU	Service Data Unit
SGSN	Serving GPRS Support Node
SLIP	Serial Line Internet Protocol
SMS	Short Message Service
SNDCP	Subnetwork Dependent Convergence Protocol
SRNS	Serving Radio Network Subsystem

TBF	Temporary Block Flow
TBF	Transport Block Format
TCH	Traffic Channel
TCH/F14.4	Traffic Channel, Full Rate with 14.4 kbit/s
TCH/F9.6	Traffic Channel, Full Rate with 9.6 kbit/s
TCP	Transmission Control Protocol
TCTF	Target Channel Type Field (in UMTS MAC)
TDMA	Time Domain Multiple Access
TE	Terminal Equipment
TLS	Transport Layer Security
TTI	Transmission Time Interval
UDP	User Datagram Protocol
UE	User Equipment
UMD	Unacknowledged Mode Data
UMTS	Universal Mobile Telecommunications System
USSD	Unstructured Supplementary Service Data
UTRA	UMTS Terrestrial Radio Access
UTRAN	UMTS Terrestrial Radio Access Network
WAE	Wireless Application Environment
WAP	Wireless Application Protocol
WDP	Wireless Datagram Protocol
WLAN	Wireless Local Area Network
WML	Wireless Markup Language
WSP	Wireless Session Protocol
WTA	Wireless Telephony Application
WTLS	Wireless Transport Layer Security
WTP	Wireless Transaction Protocol
WWW	World Wide Web
xDSL	ADSL, HDSL, VDSL (Asymmetric, High speed, Very high speed) Digital Subscriber Line

B. Glossary

3rd Generation Partnership Project (3GPP): Standardization group that specifies UMTS. This is a partnership of European (ETSI), American (T1P1, ...) and Japanese (ARIB) standardization bodies (http://www.3gpp.org/).

Automatic Repeat Request (ARQ): A protocol technique which automatically retransmits packets that had errors in transmission.

Capacity on Demand: A principle used in GSM GPRS to make the scarce radio resources available only if there is data to send. When no data is ready for transmission, no radio resources are occupied (besides control channels).

Congestion Window: The number of bytes (or packets) that can be transmitted by a ARQ protocol without waiting for an acknowledgment. This determines a *Flight of Packets*. Also called transmission window. The term *congestion window* is mainly used in TCP.

Fast Retransmission: Mechanism in TCP to retransmit unacknowledged packets immediately if three duplicate acknowledgments arrive at the TCP sender.

FDD (Frequency Domain Duplex): Uplink and downlink traffic are applied in different frequency bands separated by the frequency duplex distance. GSM and the FDD mode of UMTS (UMTS-FDD) utilize this duplex method.

Flight of Packets: The amount of data that is currently in the pipe and not yet acknowledged at the sender when operating with an ARQ protocol.

General Packet Radio Service (GPRS): This is the packet-switched data transmission extension to GSM. It has been introduced in GSM Phase 2+. Data is packet switched in the entire mobile network (core network, access network, and also the radio interface).

Gateway GPRS Support Node (GGSN): This is the gateway of the GPRS core network to external packet data networks (e.g. the Internet).

Global System for Mobile Communication (GSM): Most important digital mobile radio standard today. Standardized by ETSI.

Internet Protocol (IP): This is *the* standard network layer protocol in almost all networks including the Internet. It provides addressing and routing functionality.

Master-Slave Concept: A principle used in GSM GPRS where a single PDCH is acting as master PDCH, and all other PDCHs are slaves. Only the master PDCH carries special control channels. The slave PDCHs only carry data. So not every PDCH needs to have a control channel, which optimizes available radio resources.

Medium Access Control (MAC): Intermediate protocol layer between physical layer and link layer that controls the access of multiple logical connections on a single medium.

Mobile Station (MS): A 'Mobile Station' in GSM includes the mobile equipment itself and the Subscriber Identity Module (SIM). It is superseded in UMTS by the User Equipment (UE). In UMTS a MS is an entity which is capable of accessing UMTS services via one or more radio interfaces.

Multislot Class: The multislot class defines the capabilities of a GSM-GPRS mobile station to send and/or transmit on multiple timeslots. The number of receiving and transmitting timeslots is fully defined by the multislot class.

Non-Transparent Mode: This data transmission mode provides additional features. In GSM it is characterized by an additional Radio Link Protocol which makes the link more reliable. However, the data rate may vary.

Point to Point Protocol (PPP): A link layer protocol providing a packet transmission service to a higher layer protocol (e.g. IP). it provides parameter negotiation at the connection setup phase.

RLC-AMD (Acknowledged Mode Data, UMTS): Mode of the Radio Link Control (RLC) layer in UMTS that provides *reliable* transmission over the radio interface by means of a selective repeat ARQ protocol.

RLC-UMD (*Un*acknowledged Mode Data, UMTS): Mode of the Radio Link Control (RLC) layer in UMTS that provides only *unreliable* data transmission. Only error detection is provided.

Radio Link Protocol (RLP): This is the link layer protocol in GSM-CSD. It spans from MS to MSC and is used in non-transparent mode only. It is a selective repeat ARQ protocol providing a reliable link.

Selective Acknowledgment: Technique in ARQ protocols that retransmits *only* the missing packets, and not all others in between as well, as is done in the basic go-back-N ARQ protocol.

Serial Line Internet Protocol (SLIP): Simple link layer protocol that provides a packet transmission service for the IP network layer protocol.

Serving GPRS Support Node (SGSN): Node in the GPRS core network that interfaces the core network to the GPRS (GSM) access network for packet-switched data.

TDD (Time Domain Duplex): Uplink and downlink traffic are applied in the same frequency band, but in different time instants. The TDD mode of UMTS (UMTS-TDD) utilizes this duplex mode.

TCP/IP Header Compression: Standard compression method for combined compression of TCP and IP headers over a slow link. This is also called 'van Jacobson' header compression. A standard TCP/IP header of 20+20 bytes is compressed to typically 3–5 bytes. Compression is differential, thus only the differences to previous packet headers are transmitted. This causes problems if intermediate packets are lost.

Traffic Channel (TCH): Definition of the parameters of a logical transmission channel in GSM, GPRS and UMTS. It is characterized by the data rate, channel coding scheme, data block length, data block delay, and several other parameters.

Transmission Control Protocol (TCP): Most important transport layer protocol used today. It relies on IP as lower level network protocol. Several versions exist (TCP Tahoe, TCP Reno, TCP Vegas) with slightly different functionality. TCP provides flow control by self-clocking, and the sliding-window mechanism. It is a ARQ go-back-N protocol providing a reliable data transmission.

Transmission Window: Synonym for *congestion window*.

Transparent Mode: This data transmission mode transmits data transparently to upper layers. The upper layers just 'see' a transmission medium with constant data rate.

User Equipment (UE): A mobile equipment with one or several UMTS Subscriber Identity Module(s) [20]. It is the equivalent for the term 'Mobile Station' in GSM.

UMTS Terrestrial Radio Access Network (UTRAN): This is the completely new radio access network introduced with UMTS. The radio interface is based on CDMA. Two different modes exist: FDD and TDD duplex modes. Radio bandwidth is 5 MHz.

Universal Mobile Telecommunication System (UMTS): The new upcoming 3rd generation mobile radio standard specified by 3GPP. Early versions will reuse existing GSM and GPRS core network entities. A completely new radio access network is specified (UTRAN).

WAP (Wireless Application Protocol): A set of protocols developed by the *WAP Forum* to provide wireless Internet access independently of the underlying bearer service. WAP should be available for GSM-CSD, GSM-GPRS, UMTS and also Japanese and American mobile radio standards. The protocols involve OSI layer 4 to 7. The WAP protocols are especially designed for operation over wireless links and should work from low to high capability terminals.

WLAN (Wireless Local Area Network): A wireless broadband access system with usually higher bit rates but lower mobility than current deployed public mobile networks like GSM and also UMTS. The border between WLANs and future mobile telecommunication systems like UMTS and beyond is likely to diminish.

References

1. 3rd Generation Partnership Project. *"Architecture for an All IP network, 3GPP TR 23.922, V1.0.0"*, October 1999.
2. 3rd Generation Partnership Project. *"UMTS Core Network based on ATM Transport, DTR/SMG-UMTS 23.925, V0.2.0"*, February 1999.
3. 3rd Generation Partnership Project. *"Channel coding and multiplexing examples (Release 1999), 3G TR 25.944, V3.0.0"*, March 2000.
4. 3rd Generation Partnership Project. *"Circuit Bearer Services (BS) supported by a Public Land Mobile Network (PLMN) (Release 1999), 3G TS 22.002, V3.3.0"*, March 2000.
5. 3rd Generation Partnership Project. *"Circuit Switched Data Bearer Services (Release 1999), 3G TS 23.910, V3.0.0"*, March 2000.
6. 3rd Generation Partnership Project. *"General Packet Radio Service (GPRS); Service description; Stage 2 (Release 1999), 3G TS 23.060, V3.3.0"*, April 2000.
7. 3rd Generation Partnership Project. *"MAC Protocol Specification (Release 1999), 3G TS 25.321, V3.3.0"*, March 2000.
8. 3rd Generation Partnership Project. *"Multiplexing and channel coding (FDD) (Release 1999), 3G TS 25.212, V3.2.0"*, March 2000.
9. 3rd Generation Partnership Project. *"Multiplexing and channel coding (TDD) (Release 1999), 3G TS 25.222, V3.2.1"*, March 2000.
10. 3rd Generation Partnership Project. *"Network Architecture (Release 1999), 3G TS 23.002, V3.3.0"*, March 2000.
11. 3rd Generation Partnership Project. *"Packet Data Convergence Protocol (PDCP) (Release 1999), 3G TS 25.323, V3.1.0"*, March 2000.
12. 3rd Generation Partnership Project. *"Physical channel and mapping of transport channels onto physical channels (TDD) (Release 1999), 3G TS 25.221, V3.2.0"*, March 2000.
13. 3rd Generation Partnership Project. *"Physical Layer – General Description (Release 1999), 3G TS 25.201, V3.0.2"*, March 2000.
14. 3rd Generation Partnership Project. *"QoS Concept and Architecture (Release 1999), 3G TS 23.107, V3.2.0"*, March 2000.
15. 3rd Generation Partnership Project. *"Radio Interface Protocol Architecture (Release 1999), 3G TS 25.301, V3.4.0"*, March 2000.
16. 3rd Generation Partnership Project. *"RLC Protocol Specification (Release 1999), 3G TS 25.322, V3.2.0"*, March 2000.
17. 3rd Generation Partnership Project. *"Services provided by the Physical Layer (Release 1999), 3G TS 25.302, V3.4.0"*, March 2000.
18. 3rd Generation Partnership Project. *"UTRAN Iu Interface User Plane Protocols (Release 1999), 3G TS 25.415, V3.2.0"*, March 2000.
19. 3rd Generation Partnership Project. *"UTRAN Overall Description (Release 1999), 3G TS 25.401, V3.2.0"*, March 2000.

20. 3rd Generation Partnership Project. *"Vocabulary for 3GPP Specifications (Release 1999), 3G TR 21.905, V3.0.0"*, March 2000.
21. T. Alanko, M. Kojo, H. Laamanen, M. Liljeberg, M. Moilanen, and K. Raatikainen. "Measured Performance of Data Transmission Over Cellular Telephone Networks". Technical report, University of Helsinki, Department of Computer Science, November 1994.
22. M. Allman et al. *"RFC 2581, TCP Congestion Control"*, April 1999.
23. A. C. Auge, J. L. Magnet, and J. P. Aspas. "Window Prediction Mechanism for Improving TCP in Wireless Asymteric Links". In *Proc. IEEE Globecom'98, Australia, 8-12. November.*, 1998.
24. B. Walke. *"Datenkommunikation I, Teil 1: Verteilte Systeme, ISO/OSI–Architekturmodell und Bitbertragungsschicht"*. Dr. Alfred Hüthig Verlag, Heidelberg, 1987.
25. Y. Bai, G. Wu, and A. T. Ogielski. "TCP/RLP coordination and interprotocol signalling for wireless Internet". In *IEEE 49th Vehicular Technology Conference*, volume Vol. 3, pages 1945–1951, May 1999.
26. A. Bakre and B. Badrinath. "TCP: Indirect TCP for mobile hosts", 1994.
27. H. Balakrishnan, V. N. Padmanabhan, and R. H. Katz. "The Effects of Asymmetry on TCP Performance". In *Proc. 3rd ACM/IEEE MobiCom*, September 1997.
28. H. Balakrishnan, S. Seshan, E. Amir, and R. H. Katz. "Improving TCP/IP Performance over Wireless Networks". In *Proc. 1st ACM Int'l Conference on Mobile Computing and Networking (Mobicom)*, November 1995.
29. L. E. Baum, T. Petrie, G. Soules, and N. Weiss. "A maximazation technique occuring in the statistical analysis of probabilistic functions of Markov chains". *Ann. Math. Statist.*, Vol. 41:164–171, 1970.
30. C. Bettstetter, H.-J. Vögel, and J. Eberspächer. "GSM Phase 2+ General Packet Radio Service GPRS: Architecture, Protocols, and Air Interface". *IEEE Communications Surveys*, Vol. 2(No. 3):2–14, Third Quarter 1999.
31. S. Biaz and N. H. Vaidya. "Discriminating Congestion Losses from Wireless Losses using Inter-Arrival Times at the Receiver". In *IEEE Symposium ASSET'99, Richardson, TX, USA*, March 1999.
32. L. S. Brakmo and L. L. Petersen. "TCP Vegas: End to End Congestion Avoidance on a Global Internet". *IEEE Journal on Selected Areas in Communications*, Vol. 13(No. 8):1465–1480, October 1995.
33. B.Rathke, M.Schlger, and A.Wolisz. "Systematic Measurement of TCP Performance over Wireless LANs". Technical report, Telecommunication Networks Group, Technische Universität Berlin, December 1998.
34. K. Brown and S. Singh. "M–TCP: TCP for Mobile Cellular Networks". *ACM Computer Communications Review*, Vol. 27(No. 5):19–43, October 1997.
35. R. Caceres and L. Iftode. "Improving the Performance of Reliable Transport Protocols in Mobile Computing Environments". *IEEE Journal on Selected Areas in Communications*, Vol. 13(No. 5):850–857, June 1995.
36. J. Cai and D. J. Goodman. "General Packet Radio Service in GSM". *IEEE Communications Magazine*, pages 122–131, October 1997.
37. A. Campbell, J. Gomez, C. Y. Wan, Z. Turanyi, and A. Valko. *"Cellular IP"*, October 1999.
38. J. Case, K. McCloghrie, M. Rose, and S. Waldbusser. *"RFC 1905, Protocol Operations for Version 2 of the Simple Network Managment Protocol (SNMPv2)"*, January 1996.
39. *"The Cellular IP Project at Columbia University"*.
40. D. Cong, M. Hamlen, and C. Perkins. *"RFC 2006, The Definitions of Managed Objects for IP Mobility Support using SMIv2"*, October 1996.

41. D. Raggett et al. *"HTML 4.0 Specification, W3C Recommendation 18 December 1997, REC–HTML40–971218"*, September 1997.
42. S. Deering and R. Hinden. *"RFC 2460, Internet Protocol, Version 6 (IPv6 Specification"*, December 1998.
43. M. Degermark, B. Nordgren, and S. Pink. *"RFC 2507, IP Header Compression"*, February 1999.
44. A. P. Dempster, N. M. Laird, and D. B. Rubin. "Maximum likelihood from incomplete data via the EM algorithm". *Journal of Royal Statistic Society*, Vol. 39(No. 1):1–38, 1977.
45. A. DeSimone, M. C. Chuah, and O. C. Yue. "Throughput Performance of Transport-Layer Protocols over Wireless LANs". In *Proc. IEEE Globecom*, pages 542–549, 1993.
46. T. Dierks and C. Allen. *"RFC 2246, The TLS Protocol, Version 1.0"*, January 1999.
47. J. Eberspächer and H.-J. Vögel. *"GSM Global System for Mobile Communication"*. B. G. Teubner, 1997.
48. ERICSSON. "TEMS, User manual", 1998.
49. C. Erlandson and P. Ocklind. "WAP – The Wireless Application Protocol". *Ericsson Review*, 4(4):150–153, 1998.
50. R. Fielding et al. *"RFC 2616, Hypertext Transport Protocol – HTTP/1.1"*, June 1999.
51. European Telecommunications Standards Institute ETSI. *"GSM Technical Specification GSM 03.54 version 7.0.0 Release 1998, Description for the use of a Shared Inter Working Function (SIWF) in a GSM PLMN; Stage 2"*, August 1998.
52. European Telecommunications Standards Institute ETSI. *"GSM Technical Specification GSM 03.64 version 7.1.0 Release 1998, General Packet Radio Service (GPRS); Overall description of the GPRS radio interface; Stage 2"*, November 1998.
53. European Telecommunications Standards Institute ETSI. *"GSM Technical Specification GSM 02.34 version 7.0.0 Release 1998, High Speed Circuit Switched Data (HSCSD); Stage 1"*, August 1999.
54. European Telecommunications Standards Institute ETSI. *"GSM Technical Specification GSM 03.34 version 7.0.0 Release 1998, High Speed Circuit Switched Data (HSCSD)– Stage 2"*, August 1999.
55. European Telecommunications Standards Institute ETSI. *"GSM Technical Specification GSM 04.21 version 7.0.2 Release 1998, Rate adaptation on the Mobile Station Base Station System (MS–BSS) Interface"*, July 1999.
56. European Telecommunications Standards Institute ETSI. *"GSM Technical Specification GSM 05.08 version 7.1.1 Release 1998, Radio Subsystem Link Control"*, December 1999.
57. European Telecommunications Standards Institute ETSI. *"GSM Technical Specification GSM 08.20 version 8.1.0 Release 1998, Rate adaptation on the Base Station System – Mobile-services Switching Centre (BSS–MSC) interface"*, nov 1999.
58. European Telecommunications Standards Institute ETSI. *"GSM Technical Specification GSM 05.03 version 8.5.0 Release 1999, Channel Coding"*, July 2000.
59. European Telecommunications Standards Institute ETSI. *"GSM Technical Specification GSM 09.07 version 7.2.0 Release 1998, General requirements on interworking between the Public Land Mobile Network (PLMN) and the Integrated Services Digital Network (ISDN) or Public Switched Telephone Network (PSTN)"*, January 2000.

60. European Telecommunications Standards Institute ETSI. *"GSM Technical Specification GSM 09.60 version 7.5.0 Release 1998, General Packet Radio Service (GPRS); GPRS Tunneling Protocol (GTP) across the Gn and Gp Interface"*, August 2000.
61. European Telecommunications Standards Institute ETSI. *"GSM Technical Specification GSM 02.60 version 7.4.0 Release 1998, General Packet Radio Service (GPRS); Service Description; Stage 1"*, April 2000.
62. European Telecommunications Standards Institute ETSI. *"GSM Technical Specification GSM 03.02 version 7.1.0 Release 1998, Network architecture"*, February 2000.
63. European Telecommunications Standards Institute ETSI. *"GSM Technical Specification GSM 03.60 version 7.4.0 Release 1998, General Packet Radio Service (GPRS); Service description; Stage 2"*, April 2000.
64. European Telecommunications Standards Institute ETSI. *"GSM Technical Specification GSM 04.22 version 7.1.0 Release 1998, Radio Link Protocol (RLP) for data and telematic services on the Mobile Station – Base Station System (MS-BSS) interface and the Base Station System – Mobile–services Switching Centre (BSS-MSC) interface"*, January 2000.
65. European Telecommunications Standards Institute ETSI. *"GSM Technical Specification GSM 04.60 version 8.5.0 Release 1999, General Packet Radio Service (GPRS); Mobile Station (MS) -Base Station System (BSS) interface; Radio Link Control/ Medium Access Control (RLC/MAC) protocol"*, July 2000.
66. European Telecommunications Standards Institute ETSI. *"GSM Technical Specification GSM 04.64 version 8.4.0 Release 1999, General Packet Radio Service (GPRS); Mobile Station (MS) - Serving GPRS Support Node (SGSN); Logical Link Control (LLC) layer specification"*, March 2000.
67. European Telecommunications Standards Institute ETSI. *"GSM Technical Specification GSM 04.65 version 8.0.0 Release 1999, General Packet Radio Service (GPRS); Mobile Station (MS) - Serving GPRS Support Node (SGSN); Subnetwork Dependent Convergence Protocol (SNDCP)"*, March 2000.
68. European Telecommunications Standards Institute ETSI. *"GSM Technical Specification GSM 05.02 version 8.5.0 Release 1999, Multiplexing and multiple access on the radio path"*, July 2000.
69. European Telecommunications Standards Institute ETSI. *"GSM Technical Specification GSM 05.50 version 8.2.0 Release 1999, Background for Radio Frequency (RF) requirements"*, apr 2000.
70. F.Fitzek, B.Rathke, M.Schlger, and A.Wolisz. "Simultaneous MAC-Packet Transmission in Integrated Broadband Mobile System for TCP". In *ACTS SUMMIT 1998, Rhodos, Greece*, pages 580–586, June 1998.
71. F.Fitzek, G.Carle, and A.Wolisz. "Combining Transport Layer and Link Layer Mechanism for Transparent QoS Support of IP based Applications". In *IP Quality of Service for Wireless and Mobile Networks (IQWiM99), Aachen, Germany*, April 1999.
72. S. Floyd and V. Jacobson. "Random Early Detection Gateways for Congestion Avoidance". *IEE/ACM Transactions on Networking*, August 1993.
73. J. Franks, P. Hallam-Baker, J. Hostetler, S. Lawrence, P. Leach, A. Luotonen, and L. Stewart. *"RFC 2617, HTTP Athentication: Basic and Digest Access Authentication"*, June 1999.
74. N. Fred and N. Borenstein. *"RFC 2045, Multipurpose Internet Mail Extension (MIME) Part One: Format of Internet Message Bodies"*, November 1996.
75. E. N. Gilbert. "Capacity of a burst-noise channel". *Bell System Technical Journal*, Vol. 39:1253–1266, September 1960.

76. S. Hanks, T. Li, D. Farinacci, and P. Traina. *"RFC 1701, Generic Routing Encapsulation (GRE)"*, October 1994.

77. H. Hofstetter, A. F. Molisch, and J. Hammerschmidt. "Measurement of data transmission via the GSM 9.6 kbit/s channel". In *Proc. ICT2000, Acapulco, Mexico, Mai*, pages 30–35, 2000.

78. J. Davis II, M. Goel, C. Hylands, B. Kienhuis, E. A. Lee, J. Liu, X. Liu, L. Muliadi, S. Neuendorffer, J. Reekie, N. Smyth, J. Tsay, and Y. Xiong. "Overview of the Ptolemy Project". Technical Report UCB/ERL No. M99/37, Dept. EECS, University of California, Berkeley, CA 94720, http://ptolemy.eecs.berkeley.edu/, July 1999.

79. V. Jacobson. "Congestion Avoidance and Control". *Computer Communications Review*, Vol. 18(No. 4):314–329, August 1988.

80. V. Jacobson. "Berkeley TCP Evolution from 4.3–Tahoe to 4.3–Reno". In *Proceedings of the Eighteenth Internet Engeneering Task Force*, page 365, Vancouver, B.C., September 1990. University of Britsh Columbia.

81. V. Jacobson. *"RFC 1144, Compressing TCP/IP Headers for Low-Speed Serial Links"*, February 1990.

82. V. Jacobson, C. Leres, and S. McCanne. "tcpdump". ftp://ftp.ee.lbl.gov/tcpdump.tar.Z, 1997.

83. R. Kalden, I. Meirick, and M. Meyer. "Wireless Internet Access Based on GPRS". *IEEE Personal Communications*, pages 8–18, April 2000.

84. P. Karn and C. Partridge. "Improving Round-Trip Time Estimates in Reliable Transport Protocols". *Computer Communications Review*, Vol. 17(No. 5):2–7, August 1987.

85. S. R. Kim and C. K. Un. "Throughput Analysis for Two ARQ Schemes Using Combined Transition Matrix". *IEEE Transactions on Communications*, 40(11):1679–1683, November 1992.

86. M. Kojo, K. Raatikainen, M. Liljeberg, J. Kiiskinen, and T. Alanko. "An Efficient Transport Service for Slow Wireless Telephone Links". *IEEE Journal on Selected Areas in Communications*, Vol. 15(No. 7):1337–1348, September 1997.

87. J. Korhonen, O. Aalto, A. Gurtov, and H. Laamanen. "Measured Performance of GSM HSCSD and GPRS". In *The 2001 IEEE International Conference on Communications*, June 2001.

88. I. Kulovuori, L. Sydänheimo, and M. Kivikoski. *"Using GSM to transmit data to an information system in a dynamic environment."*.

89. M. Laubach and J. Halpern. *"RFC 2225, Classical IP and ARP over ATM"*, April 1998.

90. C. H. C. Leung, Y. Kikumoto, and S. A. Sorensen. "The Throughput Efficiency of the Go-Back-N ARQ Scheme Under Markov and Related Error Structures". *IEEE Transactions on Communications*, 36(2):231–234, February 1988.

91. S. Lin and D.J. Costello. *"Error Control Coding: Fundamentals and Applications"*. Prentice-Hall, Englewood Cliffs, NJ, USA, 1983.

92. G. Löffelmann. "TCP/IP over UMTS". Master's thesis, Institut für Nachrichtentechnik und Hochfrequenztechnik, Technische Universität Wien, Vienna, Austria, September 2000.

93. R. Ludwig, A. Konrad, and A. D. Joseph. "Optimizing the End-to-End Performance of Reliable Flows over Wireless Links". In *Fifth Annual International Conference on Mobile Computing and Networks (MobiCom '99), Seattle Washington*, pages 113–119, August 1999.

94. R. Ludwig and B. Rathonyi. "Link Layer Enhancements for TCP/IP over GSM". In *INFOCOM '99. Eighteenth Annual Joint Conference of the IEEE Computer and Communications Societies. Proc IEEE*, pages 415–422, 1999.

95. R. Ludwig, B. Rathonyi, A. Konrad, K. Oden, and A. Joseph. "Multi-Layer Tracing of TCP over a Reliable Wireless Link". In *ACM SIGMETRICS*, pages 144–154, 1999.

96. M. Failli. *"COST207 – Digital Land Mobile Radio Communications – Final Report"*. Commission of the European Communities, 1989.

97. M. Mathis, J. Mahdavi, S. Floyd, and A. Romanov. *"RFC 2018, TCP Selective Acknowledgement Options"*, October 1996.

98. G. McGregor. *"RFC 1332, The PPP Internet Protocol Control Protocol"*, May 1992.

99. D. L. Mills. *"RFC 1305, Network Time Protocol (Version 3) – Specification, Implementation and Analysis"*, March 1992.

100. "Mobile Communications, issue number 316". ISSN 0953-539x, September 2001.

101. M. Mouly and M. B. Pautet. *"The GSM system for mobile communications"*. Europe Media Duplication, 1993.

102. C. Perkins. *"RFC 2002, IP Mobility Support"*, October 1996.

103. C. Perkins. *"RFC 2003, IP Encapsulation within IP"*, October 1996.

104. C. Perkins. *"RFC 2004, Minimal Encapsulation within IP"*, October 1996.

105. C. Perkins and D. B. Johnson. *"Route Optimization in Mobile IP"*, February 2000.

106. C. E. Perkins. "Mobile IP". *IEEE Communications Magazine*, pages 84–99, May 1997.

107. P.King et al. *"Handheld Device Markup Language Specification"*, April 1997.

108. J. Postel. *"RFC 768, User Datagram Protocol"*, August 1980.

109. J. Postel. *"RFC 791, Internet Protocol"*, September 1981.

110. J. Postel. *"RFC 792, Internet Control Message Protocol"*, September 1981.

111. J. Postel. *"RFC 793, Transmission Control Protocol"*, September 1981.

112. L. R. Rabiner. "A Tutorial on Hidden Markov Models and Selected Applications in Speech Recognition". *Proceedings of the IEEE*, Vol. 77(No. 2):257–286, February 1989.

113. T. S. Rappaport. *"Wireless Communications, Principles and Practice"*. Prentice Hall, PTR, 1996.

114. K. Ratnam and I. Matta. "WTCP: An Efficient Mechanism for Improving TCP Performance over Wireless Links". In *Third IEEE Symposium on Computers and Communications, ISCC'98, Proceedings*, pages 74–78, 1998.

115. J. Rendon, F. Casadevall, D. Serarols, and J. L. Faner. "Analysis of Snoop TCP protocol in GPRS system". *Electronics Letters*, Vol. 37(No. 10):651–652, May 2001.

116. M.Y. Rhee. *"Error Correcting Coding Theory"*. McGraw-Hill, New York, 1989.

117. L. Romkey. *"RFC 1055, A nonstandard transmission of IP datagrams over serial lines: SLIP"*, June 1988.

118. W. Simpson. *"RFC 1661, The Point-to-Point Protocol (PPP)"*, July 1994.

119. J. Solomon. *"RFC 2005, Applicability Statement for IP Mobility Support"*, October 1996.

120. J. D. Solomon. *"Mobile IP, The Internet Unplugged"*. Prentice-Hall, Inc., 1998.

121. J. S. Stadler and J. Gelman. "Performance Enhancement for TCP/IP on a Satellite Channel". In *Proc. Military Communications Conference, MILCOM 98*, volume Vol. 1, pages pp. 270–276, 1998.

122. M. Stangel and V. Bharghavan. "Improving TCP Performance in Mobile Computing Environments". In *Proc. ICC'98, IEEE International Conference on Communications, Atlanta, GA, USA, 7-11 June 1998*, volume Vol. 1, pages 584–589, June 1998.

123. R. Steele. *"Mobile Radio Communications"*. Pentech Press Ltd., London, 1992.

124. R. W. Stevens. *"TCP/IP Illustrated: Volume 1, The Protocols"*. Addison-Wesly, 1994.

125. W. Turin and M. M. Sondhi. "Modeling Error Sources in Digital Channels". *IEEE Journal on selected areas in communications*, Vol. 11(No. 3):340–347, April 1993.

126. "Universal Mobile Telecommunications System (UMTS), Selection procedures for the choice of radio transmission technologies of the UMTS (UMTS 30.03 Version 3.2.0", April 1998.

127. University of California at Berkeley, Department of Electrical Engeneering and Computer Sciences. *"The Almagest: Ptolemy 0.7 User's Manual"*, April 1997.

128. A. G. Valko. "Cellular IP: A New Approach to Internet Host Mobility". *ACM Computer Communication Review*, pages 50–65, January 1999.

129. B. Walke. *"Mobilfunknetze und ihre Protokolle, Band 1"*. B. G. Teubner, 1998.

130. WAP Forum. *"Push Access Protocol Specification"*, November 1999.

131. WAP Forum. *"Push Message Specification"*, August 1999.

132. WAP Forum. *"Push OTA Protocol Specification"*, November 1999.

133. WAP Forum. *"Push Proxy Gateway Service Specification"*, August 1999.

134. WAP Forum. *"Service Indication Specification"*, November 1999.

135. WAP Forum. *"Service Loading Specification"*, November 1999.

136. WAP Forum. *"Wireless Control Message Protocol Specification, Version 4.9.1999"*, August 1999.

137. WAP Forum. *"Wireless Datagram Protocol Specification, Version 5.11.1999"*, November 1999.

138. WAP Forum. *"Wireless Identity Module, Part: Security, Version 5.11.1999"*, November 1999.

139. WAP Forum. *"Wireless Markup Language Specification, Version 1.2"*, November 1999.

140. WAP Forum. *"Wireless Session Protocol Specification, Version 11.6.1999"*, June 1999.

141. WAP Forum. *"Wireless Telephony Application Interface Specification"*, November 1999.

142. WAP Forum. *"Wireless Telephony Application Interface Specification, GSM Specific Addendum"*, November 1999.

143. WAP Forum. *"Wireless Telephony Application Specification, Version 1.1"*, November 1999.

144. WAP Forum. *"Wireless Transaction Protocol Specification, Version 11.6.1999"*, June 1999.

145. WAP Forum. *"Wireless Transport Layer Security Specification, Version 5.11.1999"*, November 1999.

146. E. Whitehead, U. C. Irvine, and M. Murata. *"RFC 2376, XML Media Types"*, July 1998.

147. J. Wigard and P. Mogensen. "A simple mapping from C/I to FER and BER for a GSM type of air-interface". In *"Seventh IEEE International Symposium on Personal, Indoor and Mobile Radio Communications (PIMRC'96)"*, volume Vol. 1, pages 78–82, 15-18 Oct. 1996.

148. A. Wolisz. *"Wireless Internet Architectures: Selected Issues"*, chapter Mobile Networks, pages 1–16. "Personal Wireless Communications". Kluver Academic Publishers, Boston/Dordrecht/London, 2000.
149. J. W. K. Wong and V. C. M. Leung. "Improving End-to-End Performance of TCP using Link-Layer Retransmissions over Mobile Internetworks.". In *Proc. IEEE Intern. Conference on Communications, ICC'99*, volume Vol. 1, pages 324–328, June 1999.
150. M. Zorzi and R. R. Rao. "Perspectives on the Impact of Error Statistics on Protocols for Wireless Networks". *IEEE Personal Commuications*, pages 32–40, October 1999.
151. M. Zorzi, R. R. Rao, and L. B. Milstein. "Error statistics in data transmission over fading channels". *IEEE Transactions on Communications*, Vol. 46:1468–77, November 1998.

Index